D0948717

Education in the Age of Science

Education in the Age of Science

edited by BRAND BLANSHARD

Essay Index Reprint Series

BOOKS FOR LIBRARIES PRESS
FREEPORT, NEW YORK

Reprinted 1971 by arrangement with
Basic Books, Inc.

INTERNATIONAL STANDARD BOOK NUMBER:
0-8369-2144-5

LIBRARY OF CONGRESS CATALOG CARD NUMBER:
70-142608

PRINTED IN THE UNITED STATES OF AMERICA

Contents

The Contributors

This book, a searching examination of American education by professionals, is the collective contribution of a group of eminent teachers and critics who have been leaders of American educational philosophy. It is based on their discussions in a seminar sponsored by the Tamiment Institute at Tamiment, Pa., in June, 1958, and also on further essays in *Daedalus,* the journal of the American Academy of Arts and Sciences, Gerald Holton, editor. All the essays here have previously been published in *Daedalus* and are used with its permission.

The contributors are:

Participants in the Tamiment Conference

BRAND BLANSHARD, Sterling Professor of Philosophy, Yale University, chairman.

SIDNEY HOOK, Professor of Philosophy, New York University.

GEORGE N. SHUSTER, President, Hunter College.

ARTHUR BESTOR, Professor of History, University of Illinois.

JOHN L. CHILDS, Professor Emeritus of Education, Teachers College, Columbia University.

REINHOLD NIEBUHR, Vice President, Union Theological Seminary.

HANS J. MORGENTHAU, Director of the Center for the Study of American Foreign and Military Policy, University of Chicago.

DOUGLAS BUSH, Gurney Professor of Literature, Harvard University.

ERNEST NAGEL, John Dewey Professor of Philosophy, Columbia University.

Discussants

JAMES E. ALLEN, Commissioner of Education, New York.

CHARLES H. BOEHM, Superintendent of Public Instruction, Pennsylvania.

SCOTT BUCHANAN, former Dean, St. John's College, Annapolis.

R. FREEMAN BUTTS, Professor of Education, Teachers College, Columbia University.

J. G. COHEN, Dean Emeritus, the College of the City of New York.

JAMES DIEKHOFF, Dean, Western Reserve University.

RICHARD H. HEINDEL, President-elect, Wagner College.

NORMAN JACOBS, Educational Director, Tamiment Institute.

BEN JOSEPHSON, JR., Nottingham University, England.

GAIL KENNEDY, Professor of Philosophy, Amherst College.

POLYKARP KUSCH, Professor of Physics, Columbia University.

ROBERT B. LINDSAY, Hazard Professor of Physics and Dean of the Graduate School, Brown University.

PRISCILLA ROBERTSON, Editor, *The Humanist.*

MELVIN M. TUMIN, Associate Professor of Sociology and Anthropology, Princeton University.

ERNEST VAN DEN HAAG, Associate Professor of Economics and Sociology, New York University.

PAUL WOODRING, Director of the Fund for the Advancement of Education, Ford Foundation.

Other Contributors

PHILIPPE LECORBEILLER, Professor of General Education and Applied Physics, Harvard University.

MARGARET MEAD, Associate Curator, American Museum of Natural History.

WARREN WEAVER, Vice President, Rockefeller Foundation.

FLETCHER G. WATSON, Professor of Education, Harvard Graduate School of Education.

DAVID RIESMAN, Henry Ford Professor of Social Sciences, Harvard University.

ALFRED NORTH WHITEHEAD, English mathematician and philosopher, 1861–1947.

Introduction

Education has had a strange history in the United States. We have never been agreed as to where we wanted it to go, so it has plunged about wildly in response to local pressures and changing fashions. For a long time it copied Europe and put its emphasis on the traditional Latin, Greek, and mathematics. Under the fire of critics as diverse as President Eliot of Harvard and John Dewey, this classical curriculum came to be regarded as an inadequate preface to life in the modern world, and for several decades American education has been trying to redesign itself. But it has failed to achieve agreement about either ends or means. What are we primarily trying to make of our students? People who are proficient in their callings? People who will have the widest knowledge of the world they are to live in? Or people of the most highly disciplined intellect and perception? One cannot have all these things at once, for the means necessary for one of them will get in the way of the means required for the others.

The recent revelation of what Russia has been achieving in science caused an already bubbling educational pot to boil a little faster. We had thought we were behind no one in our reverence for science or in our scientific achievement. We found in fact that we were lagging distinctly behind in some fields of science, that we were producing far fewer scientists and engineers than the Russians, and that there was a disquieting reluctance on the part of our high-school students to choose science as a vocation. These discoveries produced a spate of extemporized proposals, many of them ill-judged, some of them merely hysterical. They produced other responses that have been notably sane and far-seeing, particularly the report of the President's commission on science in education, which appeared in May of 1959. There is a widespread feeling that the aims of our education should be reviewed and reassessed. The chapters that follow are a response to that feeling—unofficial indeed, and made from many different points of view, but made by persons whose names and work command respect. The tale of how the book came to be written deserves to be briefly told.

The Tamiment Institute in New York carries on an imaginative program under the direction of Mr. Norman Jacobs. One of its activities is an annual seminar of experts on some topic of major current interest. In 1958 the Institute invited a group of eight leaders in educational thought to come to the Tamiment camp in the Pocono hills of Pennsylvania and spend two June days in talking over the educational problems that were on their minds. Each was instructed to prepare a paper airing some of his chief concerns, and (with a prudent spice of malice) each was paired off with someone who would probably take a divergent view. There were four sessions, at each of which two of the readers, without presenting their papers in full, gave a summary of their conclusions. Besides the readers, there were present about a dozen others, all concerned with education. There were the state commissioners of education for New York and Pennsylvania, a Nobel prizeman in physics, several professors of philosophy, education and sociology, several university deans, a college president-elect, an editor, a Fulbright fellow, and the representative of a large foundation. Their names are listed on a separate page. They all sat around a big baize-covered table, rather like conferees in Geneva, and as soon as the two openers of the session had completed their summaries, launched into criticism of what had been said. The discussion generally lasted about three hours. There was no silent moment throughout the sessions. One hardly noticed the clicking of a stenotype machine that was faithfully collecting the words as they fell.

I had the pleasant privilege of chairing these discussions, and was sentenced to pay for it by editing them later. I looked forward uneasily to the business of presiding, for the group of eight were all individuals with marked personalities and pronounced views, whose cavortings might make it difficult, I thought, to keep them running in harness. Fortunately they all proved to be sportsmen too, whose gallantry to each other and to the chair I remember gratefully. Readers of their essays who have not heard them in action will no doubt welcome a further word or two about them.

First came Sidney Hook, that inexhaustible geyser of books, lectures, and essays, a philosopher who scents the smell of battle from afar and is soon in the midst of it, giving as well as he gets, and usually somewhat better. Then there was George Shuster, the urbane head of Hunter College, whose mind, one felt, was less in controversy than somewhere above the conflict, probably on some contemplative upland where, like the poet he talks about, he could muse under his own apple tree. Next came Douglas Bush, Harvard scholar, critic, and castigator of Philistinism in all its forms. Whether mass vulgarity in this country is as dark as his portrait of it was a point on which opinion differed, but as to the skill and humor of his portrayal there could hardly be two views. He was followed by Ernest Nagel, the gentle but sharp-minded logician from Columbia, whose appeal is always to the "sovereign reason" that he thinks is found at its best in science. Then there was Arthur Bestor, the Illinois historian, lately returned from a year as Harmsworth Professor at Oxford, whose *Educational Wastelands* lit a bomb under our complacency about American schools. His position was challenged by John Childs, of Teachers

College, on behalf of that Deweyan tradition in education for which Mr. Childs is one of the most respected and persuasive spokesmen. Next came Reinhold Niebuhr, most versatile of American theologians, who was discussing a subject second only to theology in his concern, namely, America's role in the international scene, and how we are to sustain it. Unhappily, neither Mr. Niebuhr nor Mr. Childs was able to be present in person, and their statements were read for them by Norman Jacobs. The last of the symposiasts was Hans Morgenthau, political scientist of the University of Chicago, whose experience in many countries lent weight to his view that politics is in essence a pursuit of power.

It was the common feeling at the end of the sessions that the sparks that had been generated should not be allowed simply to go out. Gerald Holton, who edits the admirable quarterly *Daedalus* for the American Academy of Arts and Sciences, arranged to print the papers in slightly abbreviated form in the winter number following; and he added to them several timely essays in the same field by Margaret Mead, Philippe LeCorbeiller, Warren Weaver, Fletcher Watson, David Riesman, and the late Alfred North Whitehead. Mr. Leon Svirsky of the Basic Books house of New York thought that if all of them could be published together, supplemented by excerpts from the oral discussion, they would make a useful book. So did I. So we set to work. It must be confessed that between us we have covered a long distance with our blue pencils, chiefly, of course, on the Tamiment oral discussions. But care was always taken not to tamper with the writer's or speaker's meaning.

The first eight papers in the volume, with the discussions following each pair, present four symposia on educational problems of the day. First there is a discussion by Messrs. Hook and Shuster of the aims of education. Then Messrs. Bestor and Childs consider whether our schools are attempting too much and so failing in their intellectual task; the two men give differing answers. Third, Messrs. Niebuhr and

Morgenthau canvass the question, What can be done to prepare students for the sort of citizenship demanded by America's new place in the world? Fourth, Messrs. Bush and Nagel debate the functions and place of the humanities and the sciences in education.

What is the upshot of all these discussions? In terms of definite and agreed-upon recommendations, not very much. In terms of fresh suggestion, frank criticism, illuminating *obiter dicta,* personal enthusiasms and skepticisms, and the sort of chastening of one's views that comes from conflict with one's peers, a good deal. Certainly the discussions helped me toward a view of my own; and since in the interests of impartiality as chairman I had to stay out of the fray and try to look severely judicial, I hope I may be allowed to get into it now. I speak here for no one but myself; indeed, I should expect a sharp reprimand from some of my colleagues for the line I am about to take.

My first remark will be a fairly safe one. In the debate still raging as to whether we should stress the humanities or the sciences in education, the answer is surely: we must give strong attention to both. That dispute was already going on a century ago, as a result not of sputniks and rockets but of Darwin. When John Stuart Mill discussed the issue in his classic rectorial address at St. Andrews, he gave the answer we have just repeated. Plainly we cannot spare either the humanities or the sciences if we are to be citizens of the modern world. Granting this, should the two fields receive *equal* emphasis in our schools and colleges?

To that question I think the answer must be No. Since the Tamiment conference my opinion has been fortified by listening to a debate on this point in the University of London. Two eminent British classical scholars engaged two eminent scientists, with Sir Cyril Hinshelwood, who is at once a classical scholar and President of the Royal Society, in the chair. Since the debate was held in conjunction with the annual meeting of the British Chemical Association, the audience consisted largely of scientists. This scientific audience voted overwhelmingly that the central role in education should be assumed by the *humanities.* These people obviously did not hold science cheap; they were giving their lives to it. What did their vote mean?

It meant, I think, that science, pursued as a scientist pursues it, is a subject for the specialists, while the humanities are for everybody. By the humanities I mean such subjects as literature, languages, history, philosophy, and art. I do not think, nor did the London scientists, that these by themselves are the whole of a liberal education. Science obviously has a part in it. But at all levels in such education, its part is secondary, not primary. Why?

Because, of the two most important goals in education, science can achieve one and the humanities can achieve both. These two goals are a disciplined sense of values and a disciplined power to think. That intensive work in any science can produce a disciplined power to think is, I suppose, unquestionable. But no one who wanted a student to achieve a critical sense of what was best in poetry or music or morals would send him to physics or chemistry to get it. Darwin and Galton have left it on record that their imaginative power seemed actually to have decayed as their scientific mastery increased. By almost universal agreement the sense of moral and aesthetic value, so far as not gained by the contagion of example, is most effectively cultivated by literature and the arts, by history, biography, and ethics.

It is often contended, however, that when the humanities have done this essential work for anyone, this is virtually the end of their service; that the distinctively intellectual business of forming hypotheses and elaborating them, of testing them against fact, of sifting and weighing evidence, of proving and disproving a case, must be learned from the study of science. This I think is untrue. I do not deny that one can learn these invaluable arts from scientific study. I do deny that one cannot also learn them from the humanities, and I am inclined to think that most people learn more about them from the humanities they have studied than from the sciences they have studied. I am myself badly educated on the scientific side, and so am not a good judge. But such skill in analysis and argumentation as I have managed to pick up has certainly not come chiefly from physics or chemistry, or even from such mathematics as I was exposed to, but rather from fledgling ventures in debating, from the struggle to write essays in philosophy, and from recurrent attacks of hero-worship directed at such assorted masters of argumentation as Burke, Mill, Bradley, and Asquith.

The mind of a first-rate mathematician like Newton is a fearful and wonderful engine. The mind of a first-rate biologist like Darwin is less Olympian, but still a fine instrument. No doubt both of these men were able to carry over into the general business of life some of the precision and order with which they attacked their own problems, though Newton could write very foolishly about the prophecies of Daniel, and Darwin found Shakespeare insupportable. Still, their types of thinking are not of the sort most of us are called upon to do. Few mathematicians think mathematically when planning a vacation or voting for President, and if they waited until the abstract figures and the neat necessities of their demonstrations appeared, they would not plan or vote at all. Furthermore, as Bertrand Russell points out, much of mathematics, even higher mathematics, is hardly to be described as thinking; even the composition of *Principia Mathematica* was largely, he says, an automatic following of rules, a thinking with the fingers, as it were. Thinking in the more concrete sciences is nearer home. But here again the material, compared with that of ordinary life, is oversimplified. Things are not persons: their attributes can be isolated and dealt with by experiment; they can be measured and their relations precisely stated; they do not normally engage our passions and prejudices. Persons do. And the thinking that most men are called upon to do involves persons: it is thinking about their families, their unions, their businesses, their professions, their politics.

Now granting that the physical scientist is better at such thinking than the average man, I am inclined to think that the truly disciplined humanist is better still. Of course, if humanism means wallowing in Dylan Thomas and surrealist art, *cadit quaestio.* But by a disciplined humanist I mean, for example, a person who can read and enjoy Burke's speech on *Conciliation,* Mill's *Liberty,* Boswell's *Johnson,* and Adam Smith's *Wealth of Nations.* The man who can do that, even if he has never been in a laboratory, is in my judgment better prepared even on the intellectual side than the man trained exclusively by science. In strictly scientific problems he may be hopeless. But for the thinking he has to do as father, breadwinner, and citizen, he is much the better equipped of the two.

It is inevitable that one who takes this view should be thought to disparage science. Far from disparaging science, I think that it is increasingly important to one's understanding of the world, and that many more students should go into it professionally. But I am concerned with what should be taught to the ordinary student, and here I seem to hold an unconventional view. The study of science is not the only way to hone an edge on one's intellectual faculties. That, I have suggested, can be done as efficiently and more congenially by the humanities. What science, and only science, can do is to give us its results. The worlds it has opened, submicroscopic and supergalactic, what it has brought to light about the past of the earth and its tenants, about health and disease, about the workings of the body and the more mysterious workings of the mind—these are vastly illuminating and they should be part of the equipment of every educated person. Can they be acquired without a mastery of the techniques involved in their discovery? Scientists are continually telling us that they cannot. Is this contention sound?

The answer, I suggest, is that it is sound in the subjects which it is less essential for students to know, and unsound in the more essential ones. No one can properly understand relativity or quantum physics without a stiff dose of advanced mathematics and probably some exacting work in the laboratory. But such an understanding of relativity or quantum physics is the business of the specialist, not of the arts undergraduate. No one can fully understand the constitution of nylon or terrylene or the operation of penicillin without doing advanced work in chemistry—granted. But here again I do not see that such understanding, valuable as it is, is any necessary part of a liberal education. To be sure, the new complexion that has been given to our world by such researches in physics and chemistry *is* part of a general education. But this can be transmitted illuminatingly without an extensive mathematical and physical apparatus; it has actually been done by such expositors as Bertrand Russell and A. S. Eddington. One does not, through reading them, understand fully what went on in the minds of Rutherford, Heisenberg and company, but such understanding is neither accessible to the ordinary mind nor would it have much value if it were, in the life he is called upon to live.

On the other hand, the sciences which are most important for the average man to understand, the sciences of man and society, can be pursued profitably with a minimum of special techniques. Freud can be understood by any attentive reader. Sumner, Frazer, and Westermarck require nothing more technical than general intelligence and an easy chair; and the same may be said of Marshall's economics or Bryce's *American Commonwealth*. The method of these works can be assimilated along with their content. Biology, physiology, and geology are less accommodating, and for original work in them, no doubt laboratory work, field work, and training with instruments are essential. But the average student is not taking these subjects with the purpose of doing original work in them. And a very fair idea of both the results and the painstaking method of workers in these fields can be gained directly from what they have written; I know a philosopher whose wife read him the whole of the *Origin of Species* in their free evenings, to the pleasure and profit of both.

It is widely admitted that there is something amiss with the science teaching in our colleges; too many students are bored and alienated by it and avoid science as a vocation. Undoubtedly this is in part the students' fault. There is no royal road to the mastery of scientific technique; it takes hard work and long hours, and college is rosier and more carefree without it. But after nearly forty years of teaching in large and small American colleges, I am convinced that the apathy toward science is more than a matter of undergraduate laziness. It is largely due to a misconception on the part of scientists about the place of science in a liberal arts education. To them scientific method is the crown of the whole business: the rigor, the precision, and the beauty of it are their professional pride and joy; their mastery of it is at the base of their self-respect; if others are to think with a like precision and rigor, they must achieve a similar mastery of it. The experts and specialists devise courses calculated to produce experts and specialists like themselves. The few among their students who are already committed to science and proficient in it revel in these courses and get A's in them. The others, who are the great majority, fall into two classes—those for whom this adventure in science is terminal, and those who are feeling their way, wondering whether to go farther or not. For neither class do present courses hold much appeal. For the arts student they seem full of details that are unimportant and techniques that he will never use. The student who is uncommitted says quietly to himself that if science means a lifetime of this manipulation of x's and y's in the interest of ends that are somehow never made clear, then thanks very much; he has had enough; his vocation seems to lie elsewhere.

Science teachers throw up their hands at this attitude and complain of student defeatism. To them it looks as if the students are sidling off into the humanities because they cannot face rigorous requirements and high standards. I have become increasingly skeptical of this explanation as I have listened to it over the years. The American college student is, by and large, an ambitious and serious fellow, with a keen nose for the areas in which he will find intellectual profit, and it is incredible that he should go on deprecating and depreciating "the most valuable of college disciplines" merely because it is hard. If he grumbles, the reason is quite simply that introductory science, taught with its present emphasis, seems to him an investment of low return. In the contention that science is the only road to clear and accurate thinking, he smells an aroma of professional humbug. Old X in history is an ogre, but listen to his lectures for a term and you somehow understand what international issues are about; Y teaches philosophy, which everyone knows is hokum, but after he has ripped a few of your papers to pieces, you begin to know what self-critical writing means. Students collect around these people like moths around a lamp. They would collect in the same way around any science whatever if the educational candle power were there.

I do not think it will be there until scientists learn the difference between science for the liberal-arts man and science for the specialist. It must be said bluntly that failure to learn that difference is the prime reason for the plight of science in our colleges and universities. What is needed are teachers in physics, in chemistry, in biology, who are philosophers and humanists in the sense that they feel in their own minds, and can communicate to others, the importance of their subjects in understanding the modern world. A young man who has just written a Ph.D. dissertation bristling with professional jargon is not the most likely candidate for this difficult office. In my own department at Yale, we consider freshmen too hard an assignment for the younger men and throw into the task our senior professors who have worked at the subject for enough decades to have achieved some simplicity of view. Older heads are better than younger ones at distinguishing the woods from the trees. But there is no rule about this; some young men are born into swaddling clothes that take the shape of a teacher's mantle, and some scientists of world reputation are hopeless in the classroom. I once heard a man say that there was only one worse teacher of physics in Europe than Kelvin at Glasgow, namely, Helmholtz at Berlin—and the speaker had sat under both.

Above all things in the world we in America need an educated citizenry. That means, I have suggested, a citizenry with a disciplined sense of values and a disciplined power to think. Both types of discipline can be supplied by the humanities, and it is through the humanities that most people must continue to gain them. The second kind of discipline can be gained from science also. But current scientific teaching is actually preventing students from getting it by trying to make them leap over mounds of detail and technological hurdles when they are hardly clear what the enterprise is all about. There is small point in prodding and switching them over these hurdles; most of them are not cut out for that sort of mental athletics. What they need is introductions to science taught by dedicated men who will give them some conception of the achievements, the importance, and the attraction of the specific branches of inquiry, and at the same time awake in those of them who are born to the purple the excitement of a high calling. The thing is not impossible because it has been done. It was done by Nathaniel Shaler and Louis Agassiz at Harvard and by T. H. Huxley in London; indeed it is always being done by unsung scattered teachers who live in their students' memories. If we could sow such teachers two or three to a college, we should not have to worry about the future crop of American scientists.

BRAND BLANSHARD

Sidney Hook

George N. Shuster

and Panel

I

What Is Education?

SIDNEY HOOK

Fear and panic, resulting from recent technological and military advances in the Soviet Union's power, have produced outcries against our schools and spawned proposals to reorganize the whole of American education. But it is hard to find among these criticisms any items of agreement other than the familiar complaints that American education has become soft and intellectually contemptible; that professors of education are responsible for our deplorable condition (not the much larger number of professors of other subjects who until recently hardly recognized their role as educators); and that John Dewey is the evil genius who misled them. The writings of Dewey themselves remain largely unread and misread.

It is certainly true that we have not given our students—or our statesmen—an education which fits them to rise to the occasions and challenges of our time of troubles. But simple justice requires us to recognize that it was Dewey rather than any of his multitudinous critics who foresaw the general nature of our crisis. It was Dewey who called upon educators first to take note of the vast effects of the scientific revolution upon our society and its educational needs, and later to recognize the Communist threat to the free world.

In this essay I shall restate and defend what I believe the ends of American education should be for our time and our place in history. It is my thesis that, given the kind of world we live in, our society and history, the powers of the human mind and body, certain ends of education are to be justified, not by deduction from metaphysics or theology, but in terms of their fruits in experience and in the light of undisputed moral values.

For purposes of convenience I divide the ends of education into three overlapping groups: (a) powers and skills, (b) knowledge of subject matter or fields of study, and (c) moral habits, values, and loyalties.

Powers and Skills

Education should aim to develop students' capacities to write and speak clearly and effectively, to deal competently with number and figure, to think critically and constructively, to judge discriminatingly and observe carefully, to appreciate and respect personal and cultural differences, to enjoy with trained sensibility the worlds of art and music, and to enrich the imagination and deepen insight into the hearts of men by the study of literature, drama, and poetry.

Why? Many reasons may be given, but three generic reasons are sufficient. We must all communicate with each other, no matter what our business and vocation, in a world of increasing specialization. The effective exercise of these powers enables us to make our experiences more significant and to share them, if we so desire, more readily with others. Man is born to problems and troubles as the sparks fly upward. The ability to think increases the power to solve problems and increases our satisfaction in doing so. By increasing both power and satisfaction, it multiplies alternatives of choice and makes us freer men. The development of our capacities of aesthetic appreciation and imaginative identification multiplies the occasions for joy and delight in a tragic world. It refreshes the spirit without imposing suffering on other human beings.

The values in the light of which these consequences are appraised are not themselves beyond question and dispute. We may believe we can ground them in ultimate intuition, as does G. E. Moore. Or we may agree with John Dewey that they can be sustained from case to case, from problem to problem, in an unending continuum of experience. In either case, these values recommend themselves to us more validly than any allegedly justifying metaphysical or theological principle.

More controversial is the question whether the powers and skills we seek to develop in education should also include those necessary to earn a living. The question here is whether the schools should take the responsibility of helping students in their choice of and preparation for a calling, or whether industry should do it on the job; if the schools should assume this task, whether they should do it in special vocational institutions or in those of general or liberal studies; and if in the latter, at what point in schooling and with what relationship to nonvocational studies. These questions I forego for the moment except to point out that the answers depend not so much upon first philosophical principles as upon the character of the society in which our students will live and upon our conception of what constitutes a democratic education.

Fields of Study

With respect to subject matters and fields of interest, all students should acquire an adequate knowledge of the physical and biological world—of the forces that play upon and govern man's habitat, limit his place in nature, and determine the structure and behavior of his body and mind.

Why? Again for many reasons. Such knowledge is necessary to make the student's everyday experience intelligible to him. He will be more at home in the world. He will have a better understanding of scientific method in action—especially if he is properly taught. He will become more acutely aware of the revolutionary impact of science and technology on human culture. If he rises above the level of the earthworm and wonders about human origins and destinies, this knowledge will help him to develop a reflective view of the place of man in the universe, of God's existence, of the meanings and evidences, if any, of immortality and human freedom.

A second field of interest and subject matter is history and the social studies. There is a universal need for all individuals to understand the society in which they live. Every student is a future citizen who cannot make intelligent choices in political affairs, or even in some of his personal affairs, without learning something of the massive economic and social forces—whether it be the presence or absence of oil and rubber, the surge of nationalism, or the pressure of population—that mold contemporary civilization. Whether or not wars begin in the minds of men, ideological differences may well determine whether these differences result in conflict. Educators disagree not about the desirability of instruction in these subject matters but only on the relative emphasis to be placed on the distant past, the recent past, and the contemporary. It seems to me that the key to wise selection of materials from the past is *relevance* to the great issues, problems, and challenges of our age that must be mastered if we are to survive as a free culture. For example, it is far more important to study the nature of communism than the War of the Spanish Succession or even the history of Rome, although in the course of the study of communism we may find illuminating the study of some aspects of Roman history.

The fact that some selection must be made indicates that here and elsewhere we must be guided by some notion of importance, relevance, or strategic perspective. Not everything is relevant to everything; although all subject matters and all experiences have some worth to someone, they do not all have the same quality or the same worth in the educational enterprise, which seeks to make the individual feel at home in a world of change—mastering events by understanding and action instead of being altogether mastered by events, growing in such a way as not to obstruct further growth in insight and maturity.

What John Dewey says of experience seems to me also to be true of the subject matters experienced—formal grammar is not as good as poetry if we want significant experience, stamp-collecting is not as good as geography and history if we want to understand the map of the world today, the study of Greek and Latin words in English use is not equivalent to the study of good English usage for purposes of better writing and speaking. As Dewey observed in *Experience and Education*:

> The belief that all genuine education comes about through experience does not mean that all experiences are genuinely or equally educational. Experience and education cannot be directly equated to each other. For some experiences are miseducative. Any experience is miseducative that has the effect of arresting or distorting the growth of further experience. An experience may be such as to engender callousness: it may produce lack of sensitivity and responsiveness. Then the possibilities of having richer experience in the future are restricted.

A third field of study that should be required of all students, particularly in the colleges, is a study of the great maps of life. The value judgments and commitments of the major philosophies and religions that have swayed multitudes, as well as the visions of solitary figures of deeper thought but lesser influence, bear directly upon some of the ideological conflicts of our age. No wise policy can be formulated independently of the facts, but no mere recital of the facts determines policy. In the end, a decision between conflicting social philosophies involves a choice among key moral values. Sometimes this is true in what appear to be merely questions of limited means. For example, hard decisions about nuclear defense in the grim days ahead of us involve commitment to ideals on which we may literally have to stake our lives. Surely this is justification enough to make a critical and searching study of the grounds, alternatives, and consequences of the great ideals for which men have lived and died. It would be interesting to know how many of the brainwashed American prisoners of war in Korea were ever excited by a course of study in the meaning and history of human freedom.

Values and Loyalties

This last reference is a proper transition point to the aims of education that bear upon moral habits, fundamental loyalties, and what is called character education. I do not believe that the intellectual and moral virtues—whether a love of the truth, a sense of chivalry and fair play, a feeling of outrage before cruelty, sympathy for the underdog, or a passion for freedom—can be instilled by didactic instruction. They can be imparted, if at all, only by indirection, by skilled teaching on the part of teachers who care, and only when students learn well the other things encompassed by our ends. How, for example, do we go about developing intellectual and emotional maturity in students? Not by preaching but by setting them tasks of progressive complexity. If the sign of maturity is the possession of habits of reasonable expectation, I do not know how this can be built up except by getting students to learn from lesson to lesson what the world is and what it might be, and relating the possible ideal fulfillments to the limiting conditions that govern men and things. Immaturity may be as much present when we settle for too little, blind to what may be, as when we demand too much, blind to what cannot be.

It is our faith in the educational process as a whole that sustains us in our belief that those who complete their schooling will have acquired loyalties to the enduring values of the human community. Despite those who misunderstand him—and they are found equally among those who believe themselves to be followers as well as among his critics—John Dewey has placed great emphasis upon this aspect of the continuity between past and present. In *A Common Faith* he writes:

> The things in civilization we most prize are not of ourselves. They exist by grace of the doings and sufferings of the continuous human community in which we are a link. Ours is the responsibility of conserving, transmitting, rectifying and expanding the heritage of values we have received that those who come after us may receive it more solid and secure, more widely accessible, and more generously shared than we have received it.

The nature of education is such that even when learning is a process of discovery the greatest weight must fall upon the knowledge and wisdom of the past. It could not be otherwise. It takes time for the individual to discover that there are many pasts or many interpretations of the past, and that anything that is genuine knowledge must prove itself in the present, and therefore need not fear challenge. It takes more than time. It takes intellectual courage, the rarest of all intellectual virtues. Those who make a fetish of the past, of historical continuity, of piety before the traditional, live off the intellectual capital of their ancestors' courage. The gabble in the academies about the vice of conformism and the virtue of nonconformism is empty and meaningless. Hitler was the greatest nonconformist of the twentieth century. What we must cherish is not conformity or nonconformity, agreement or disagreement, but intellectual independence, the courage to hold a position, on the strength of evidence, no matter what the baying of the crowd.

Once we accept these objectives as the ends of education, or any equivalent set, I believe we can easily show that many of the antitheses that plague current discussions of the subject may be resolved. I wish to consider briefly two of them.

The Intellect Versus the Whole Man

The first is the dispute over whether the end of education should be education of the intellect or of the whole person. Both positions seem to me untenable. The intellect or mind is not an abstract, disembodied power. It influences and is influenced by our emotions. It guides perception and is checked by perception. On the other hand, although all aspects of body and mind in a person are somehow related in a pattern of personality behavior, they do not all seem equally important in determining the characteristic Gestalt. They cannot all be developed and certainly not at the same time. Except when we encounter a Leonardo, we cannot avoid selecting and developing some powers at the cost of others. The concert pianist is not likely to be in a position to cultivate his skill as a pugilist, and for reasons other than his fear about the use of his hands. There simply is not enough time to develop all of our intellectual interests, not to speak of all our practical aptitudes. Development opens up new possibilities, but it is also true that it takes place through successive limitations of possibilities. If the development of the powers of cooking, fishing, and roller skating get in the way of the development of the powers of reading, writing, and problem-solving, then the first must yield. So much for the development of the whole man.

Let us look at the mind in action. What does it mean, educationally speaking, to develop the student's powers of thinking in the biological sciences? Anyone who sets out to teach his students to think in these fields is teaching them at the same time how to see, how to observe, how to use instruments, how to discipline impatience, how to curb the impulse to take short cuts. Is all this part of the mind? Trained observation in every field is an art. It is not merely looking, for it is guided by general ideas that structure the field of perception. Thinking about machines involves knowing how to make things. Thinking is not *merely* reasoning. Otherwise we would have to regard every paranoiac as a thoughtful man. It is not accidental that thoughtful and sensible are closely related. Chesterton once remarked that pure logic was the only thing an insane man had left.

What does it mean to think about a play, or about a poem, or about people? It means also to feel, to imagine, to conjure up a vision. Why is it that we often say to some thoughtless person, "Put yourself in his place"? To another, "You haven't got the feel or the hang of it"? To a third, "You understand everything about the situation except what really matters"? We do not convey truths by this way of speaking, but we help others to find the truth. If artists and musicians think, their sensory discriminations must be relevant to the thinking they do. After all, we do speak of educated tastes. It is absurd, therefore, to say that the exclusive preoccupation of education should be the development or training of the mind.

Nonetheless, although the antithesis between these two points of view must be rejected, the accent has to fall on one rather than the other. To avoid the implicit faculty psychology associated with the term "mind," I prefer the term "intelligence." Intelligence suggests more than mere ratiocination. It suggests the ability to look for evidence and to discern the likely places where it can be found, and the capacity to weigh it judiciously. The intelligent man knows when it is time to stop reasoning and to act; when it is time to stop experimenting and to declare his results. Of him one never says that he is educated beyond his capacities. He is wise rather than learned because he knows the uses and limits of learning.

Content Versus Method

The second antithesis I wish to challenge is the one usually drawn between content and method in education. Shall one cover a great deal of ground or study in depth, stock the mind with useful information or enable it to find the facts quickly? The danger of emphasizing content rather than method in education is that unless content appears live and meaningful to students it is transformed into a dull inventory of facts. To be live and meaningful, content must be related and connected to other content, to problems and issues, and wherever possible to live options. What better way is there of establishing these connections and relations than to show the methods by which conclusions are reached? Where content is stressed at the cost of method, memory rather than understanding becomes the chief aid to educational progress. A good memory is a blessing. But even if it does not become atrophied in the world of automation, it seems as if anything a human being can remember a machine can remember better. Moreover, the power of memory alone cannot put two memories together to discover something new.

The danger of stressing method over content is said to be equally great. It sacrifices content: the student is not required to *know* anything so long as he can talk suggestively about it or around it. This emphasis tends to regard any subject matter as the equal of any other for purposes of developing proper habits of thinking. In some institutions stress on method has often led only to talk about how to talk, to the use of a sophistic dialectic, at which Plato poked fun as the infantile sickness of adolescent philosophers.

I confess that I do not see much danger to learning in giving the prime emphasis to method. By emphasis upon method, I mean upon *critical* method—upon criteria of evidence, norms of validity, rules of consistency, on "how we actually think" and "how we ought to think" in whatever field of study we want students to be informed about. This seems to me more important than stressing what we think, because the how and the what, although distinguishable, are actually inseparable when we think soundly. When facts come into dispute or inferences are questioned, we find ourselves relying on rules or habits that control observation and the movements from statement to statement. In my experience, the most critical thinkers I have known —for example, my first teacher in philosophy, Morris R. Cohen—have been the best informed. The citing of counterinstances is a phase of critical thinking. Nor does it follow that, because critical thinking should receive the greatest emphasis, any subject is as good educationally as any other for that purpose. To be sure, we can think critically about horse racing and show when it is wise to suspend judgment or hedge a bet or distrust a bookie. But it is possible to learn the same critical lessons by studying subject matter that has a wider range of generality. In some fields—for example, learning languages—I grant that memory is more helpful than critical thought. Finally, those for whom critical method is a kind of verbal sport, a dialectic by which fact is dissolved or the worse made to appear the better cause, have not learned to think critically about language or honestly about purposes. There are stubborn cases in which adolescents are more in need of psychiatry than of philosophy.

The reader will probably disagree with some of the ends of education I have here enumerated or with my relative emphasis. If he does, it will be because of some feature of human experience he believes I have overlooked. He may find some personal or social needs less pressing than I do. He may believe that others are neglected. My contention is that none of his reasonable criticisms will follow uniquely from any theory of ultimate reality.

I now present certain considerations that follow from a democratic standpoint in education. The phrase "democracy in education" has meant all sorts of bizarre things. To some people it apparently means that everything is settled by a majority vote by children in the classroom—not only what to study but what to believe about what is studied. I mean by the phrase simply this: the right of every child to equality of educational opportunity. But this is not so simple. Literally construed, it would mean that we would have to revolutionize our society to establish greater economic equality, for in the homes of the poor and the rich equal educational and cultural opportunities cannot be found. It would mean, as only Plato had the courage to see, the abolition of the family, so that all children could be brought up by the best possible educational foster parents. Soberly but not literally interpreted, democracy in education is equality of opportunity to achieve through schooling an education commensurate with one's capacities. This is America's contribution to the history of education. It requires that we grant to our neighbor's children, no matter what their social status, the same rights to an education that we demand as parents for our own children. Anyone who accepts this principle seriously must acknowledge the great responsibility of the state, as the public agency, to equalize opportunities.

The Meaning of Equality in Education

To say that all children have the same right to an education is not to say that all have the right to the same education. It does not mean that they have the right independently of their capacities to attend the same schools. It does mean some education for all. It leaves open how much and how long. It is as absurd to say that some education for all means education for none as to say that nourishment or health for all means nourishment and health for none. It would be absurd even if we confused "some education for all" with "the same education for all."

No matter how generous our hopes for mankind, to be reasonable they must be compatible with the facts of biological variation. Do the facts of biological variation defeat our ideal of democratic education, in the sense that it is futile to expect most students to profit by an education defined by the ends we have previously derived? If they do, then, as Jefferson foresaw, the prospects of our survival as a political democracy are extremely dubious. It seems to me, however, that it is quite reasonable to recognize the facts of biological variation in human capacities and still defend democracy in education. It requires that we distinguish between the *function* of schooling and the *content* of schooling, and strive to achieve the same function with different content, where content refers not to ends but to courses and methods and materials of study. These *may* be the same; they do not have to be.

It is not necessary to choose between the view that everyone should be educated in the same way and the view that liberal education is for a small elite while the rest of mankind are to be trained as hewers of wood and drawers of water for their intellectual betters. We can put this to a test by describing a hypothetical situation. Two highly gifted parents, who have achieved academic distinction although they were brought up in underprivileged homes, confirm the Mendelian laws of heredity and rear a family of children whose native intelligence ranges from very dull to very bright. Let us assume that the parents themselves undertake to teach their children, all of whom are equally dear to them. Would they not try to realize the same educational ends for *all* of them? Would they not want all their children to learn to speak and write clearly, to read and think effectively, to enjoy music and painting and the other arts of civilization, but each to the best of his ability and therefore in different measure? Where children's health is concerned, parents naturally provide special medical treatment for the weakest. Where intelligence is concerned, they naturally provide special educational opportunities for the brightest. But they are equally concerned for the health and education of all of them. If they were not, they would be bad parents. A democratic society stands in the same relation to all the children of the community as good parents stand to their own children.

Nonetheless, equality of educational concern on the part of our hypothetical parents would not necessarily lead them to give the same instruction in all subjects to all their children even if in varying amounts. What a child cannot grasp about a foreign culture by mastery of its language he may learn by reading books on travel or anthropology. There is nothing undemocratic in diversifying the courses of instruction, the rate of instruction, and the methods of instruction.

The Appeal to Experience

This brings me to the necessity of an experimental approach. In one sense, an appraisal of any proposed educational end in terms of the consequences of pursuing it is experimental. Such an approach should not be regarded with hostility by those who stress "eternal and perennial values," because if the latter are not grandiose terms, concealing some parochial or partial interest, their validity will be established in the here and now of experience.

Even if an educator claimed that his educational aims were authenticated by an infallible insight, surely he could not reasonably claim to know the methods and means by which they could best be realized. The intellectual scandal of much recent discussion between traditionalists and some progressive educators is the attempt by both sides but especially by the former to settle questions in this area not by inquiry or experiment but by dogma.

The phrase "progressive education" is today very much at a discount—and deservedly so—because of the number of educators who thought they could remain progressive while ceasing to be liberal. Originally, however, all it meant was an acceptance of the principle of democratic education and a reliance upon the findings of scientific psychology about the learning process. These two positions were revolutionary at the time they were formulated, and they still constitute the law and the prophets for modern educators, everything else being commentary. Some progressive educators have deduced what procedure should be followed in educating the young rather than following the lead of experimental evidence. But some of their critics, who have counted only the failures of progressive methods and not their successes, have been even more dogmatic and undiscriminating in their claims, holding them responsible for educational phenomena and conditions that must be laid at the door of society. There are critics who tell us that the schools have failed to teach their charges, and failed most miserably with the gifted, and in the same breath concede that their college students as a whole are more serious, abler, more excited by ideas than their precursors in the golden age that existed before the days of progressive education.

There is a great deal of what is called "experimentation" always going on in American education, but most of it is not experimental, since it is conducted without proper controls. The result is that we think we know more than we actually do about the best courses to teach and the best ways of teaching them. The fact that something is new does not make it experimental. Nor does the desirability of experimentation mean that we must keep on experimenting about the same things. By this time we already should know what are the best methods of teaching children to read, of teaching chemistry in high schools, of teaching economics in colleges.

Our very metaphors sometimes betray that we are taking for granted what might very well be in dispute. One of our leading traditionalists, Mortimer Adler, has written: "Human differences in capacity for education can be thought of in terms of containers of different sizes. Obviously a half-pint jar cannot hold as much liquid as a quart or gallon jar. Now the poorly endowed child is like the half-pint jar, and the gifted child like the quart or gallon container." He concludes not only that each container must be filled up to the brim but that each must be filled with the same rich, thick "cream of liberal education."

The comparing of children to different measures, coupled with the conception of teaching as the *pouring* of the same stuff into passive containers, expresses a point of view which is hard to reconcile with what we know about children as organic creatures and learners whose differential responses determine how much they can assimilate. Even cream cannot be poured into all children with safety, no matter how ingenious our funnels. It would help to change our metaphors. Our experimental task is to find and offer the appropriate curricular nourishment for different types of organisms that will enable them to achieve the full measure of their growth and health. That curricular nourishment may be the same or different. The test is the *function* it performs in the life of the child. The same function will not give us the same result. A dull child will never be able to read as well or as intelligently as a bright child, and an ordinary child will never be able to play as well as a musically gifted child. But both children can be so educated that each enjoys reading and music. Both can acquire something of the grace and taste associated with the liberal arts. They may not both be able to do so by studying the same subjects, even though the study of some subjects will be common to them.

"One-fourth of a Nation"

How great is the range of student capacities? *The New York Times* (4 May, 1958) summarized a report issued by the United States Army, which recruits from all classes of the population. It stated that 25 per cent of the inductees who have passed their physical tests of fitness lack the capacities to be trained for anything except simple manual tasks like cleaning, polishing, digging, and driving. A modern army requires many technical skills. It has a use for individuals with a flair for mathematics, or a knack for building machines, or a feeling for foreign languages. There is no reason to doubt the willingness of the military authorities to recognize the presence of superior and average intelligence and/or aptitudes in the ranks, or to doubt the substantial accuracy of the reports of their distribution. It is fairly safe to extrapolate the ratios to the rest of the population.

This means that approximately one-quarter of all our students are incapable of completing the requirements of a good academic high school and going on to a liberal-arts college. By increasing the number of schools and teachers, decreasing the size of classes, and improving skills of instruction, we can do something to bring down this number. Even so, the evidence shows that there will always be a large group unable to profit by the traditional and conventional courses of study. What shall we do with them? Unless they have private tutors it is they who during the period of adolescence obstruct the learning of the more gifted students. Before we answer this question, let us imagine once more, as difficult as it may be, that our own child is among them.

It seems only common sense to say that the education of this fourth of the nation should not be so organized as to interfere with or dilute the education of the other three quarters. Conversely, the education of the latter should not be a ground to deprive the former of their educational rights. The problems here are admittedly difficult. They have not been solved. But they are not insuperable. We should keep these students in school as long as they can profit significantly by instruction. We should put them in special classes if they can learn better that way. We should instruct them in the skills and subjects that will enable them to begin their vocational experience at an earlier age than their more gifted brothers and sisters—who, sooner or later, must prepare themselves to earn a living too.

Some Principles of Vocational Education

This introduces the complex problem of vocational education, which is often bedeviled by the assumption that where it begins, liberal education must end. Yet we do not make the same assumption about professional education, which is distinguished from vocational education not only because it requires better brains but because it enjoys a higher social status and more money. Until the necessity for earning a living disappears, there can be no reasonable objection to using the schools to prepare people for a good living as well as for a good life. The pity of it is that the vocational schools are so bad—worse from the point of view of their own purposes than the academic and general high schools. The reason is partly the confusion between job training and vocational education, and partly the use of vocational schools to provide occupational therapy or temporary shelter to juvenile delinquents who do not belong in schools but in corrective institutions. (These delinquents may be victims of society, and are entitled to special educational, even custodial, care, but what is educationally relevant is that at the moment they are agents of educational chaos.)

The important points about vocational education seem to me these: First, it should include those fields of study which are so general or liberal in character that they must pervade all kinds of education. They bear directly on every student's social and political responsibilities, no matter what his vocation, especially in a community where each man's vote counts for one and no more than one. Instruction in these areas must be given at every level. The student should be made constantly aware that his vocational studies may become a means of dehumanization if no thought is taken of the social contexts and moral implications in which vocational choices are often made. Second, instruction in vocational subjects should cover the basic principles that govern a whole class of practical skills for which the individual displays a bent or interest. Third, we must avoid the invidious social distinctions that accompany educational differentiation; we must recapture the sense, rapidly being lost, of the dignity of useful work. Wherever there is a stigma attached to vocations, it will be attached to education for those vocations. It will be difficult for students themselves to believe that educational segregation for certain vocational purposes, whether in the same or in different schools, is compatible with educational democracy. The doctrine of separate and equal facilities has no justification whatsoever where race is concerned, but where we segregate solely on the basis of intellectual capacity or interest, there is no reason to cry "Havoc!" We must forbear, however, from exacerbating the sting of natural resentment at being intellectually underendowed in a world where survival and power depend more and more upon brains.

Secularism and Clericalism in Education

What are the ends of education from a secular point of view? The simplest but not the clearest answer is—to keep private altars out of public schools. As I understand secularism, its opposite is not religion, because sometimes secularism itself is characterized as a religion. The term "religious" has become so ambiguous that one no longer knows where to look for the irreligious. In addition, there are conventionally religious individuals who are firmly convinced that religion is a private matter and therefore does not belong in the schools. What secularism is really opposed to is clericalism. I shall define clericalism as the belief that the acceptance of certain doctrines about the supernatural is required in order to reach truths about man, nature, and society, in order to discover morally valid ideals, and in order to acquire certain intellectual skills. The ends of education I have previously enumerated roughly fall within this threefold classification of truths, values, and skills.

What have beliefs in the supernatural got to do with the acquisition of skills? No plausible connection has been established between them.

What truths about the world, society, and man depend upon the prior acceptance of religious truth? It may be argued that there is a Christian metaphysics, but who in this age of modern science can show that its acceptance is a necessary condition of belief in a Christian physics or just a physics? Religion has withdrawn from conflict with science by renouncing any pretensions to speak about matters of fact in the dimension of nature. If it resumes the conflict, it must submit its claims to truth to the same arbitrament of method as all other scientific hypotheses. There remains, then, the claim that without some transcendental belief as a supporting or justifying ground, moral values cannot be sustained. Karl Heim, the Tübingen theologian, goes so far as to assert that the root choice of man is between secularism and Theism, and that "secularism, to be consistent, must entirely reject all such words as God, eternity, conscience, ethics, moral rearmament, guilt, responsibility . . . as inadmissible borrowings from a view of the world opposed to its own." However, no proof is given that moral judgment rests upon any theological belief. On the contrary, it is demonstrable that if any moral attributes are predicated of supernatural powers or entities, they are derived from the autonomous judgments of men.

Men always have built and always will build gods in their own moral image.

This fact in no way precludes the intensive examination of religion in the course of study. On the contrary, neither past history nor present society can be understood without an intensive analysis of the role of religious movements, traditions, doctrines, wars, and conflicts. To the extent that religion, in Santayana's phrase, "is an imaginative echo of things natural and moral in human experience," we can appreciate it as poetry. To the extent that religion makes claim to a distinctive knowledge, we can consider it in our study of philosophy. For educational purposes, is it not sufficient to approach religion as a creation of the human spirit without assuming that it is a revelation of the Divine Spirit? What more can reasonably be required without introducing into the schools of the nation the strife of private faiths? And this at a time when our more intimate contact with other cultures reinforces the wisdom of an unsectarian approach to the religions of the world?

Those who believe that religion can serve as a unifying principle around which to synthesize the subject matter of education seem to me to ignore both the nature of religion and the pluralistic commitments of the American tradition. I have more sympathy with those who see in religion an avenue to an experience that sustains human allegiance to ideal ends, especially when these are threatened with defeat. Religious experience may offer support to human ideals, but it is not the sole source of such ideals nor can it ever be the guarantee of their validity. The educational experience itself, when teachers have both skill and vision, may become both source and support of the love of truth, the love of justice, the love of beauty, and the love of human freedom. This is all the public religion we need, in both peace and war. For it enables us to hope without illusion, to fight without despair, and to stake our life in defense of the things that make life worth living.

GEORGE N. SHUSTER

What is education? We should consider it as a process before we attempt to define its concern with truth.

The schools constitute a kind of arc, the extremities of which are rooted in wholly disparate functions. At the outset the teacher is a person who tells children what it is considered desirable they should know—verbal symbols, the multiplication table, the names of rivers and seas, and phrases expressing civic, ethical, and religious beliefs. A little later there will be consideration of what Paul Weiss calls "the mastery of techniques"—of diction and reckoning, of accuracy, of the progression of thought from data to conclusions. At the other end of the academic span there is, however, in principle no concern with the imparting of knowledge. The scholar in his study, the monk in his cell, the scientist in his laboratory—if you will, the poet under his tree—for all of these the dialogue is starkly between the self and Reality. To make such conversation possible in terms of scholarship may well be the central assignment of the university.

Between the grade school and the research institute lie the reaches of education in which there takes place a sort of fusion between exploration shared and knowledge imparted. The reputable college, for example, must at least upon occasion be akin to Augustine's Cassiciacum, where in goodly fellowship problems like that of "the happy life" were discussed in the give and take of dialogue. Yet even the best of such institutions will normally be busy with things thought rather than with man thinking. Carlyle held that one must be content with enough happiness to get one's work done. Manifestly, education in its intermediate stages keeps busy giving young people sufficient knowledge (and it may be insufficient wisdom) to perform useful service in the world. Catering to utility, as a matter of fact, may tempt the fully academic mind to derisory or even ribald comment. But if all of us on campuses are quite honest, will we not admit that a great deal of what we do is related to utility in character and purpose?

At any rate, scrutiny will reveal how closely the three concerns—knowledge, inquiry, and usefulness—are intertwined. A great many young women, for instance, are trained to teach in nursery and grade schools. I think it quite probable that my own college, while little more than a secondary school, was graduating teachers fully as able to impart the kind of instruction needed as it is now doing when it has become a sedate and rather exacting college of liberal arts. Why not? If your task is to teach addition and subtraction, you need to know about these and not about the calculus, which is in the course of study only because it is believed to give the student insight into an aspect of reality that she will then know about but not use. Even more notable is the fact that we have added to the training program a great deal of information about theoretical and applied psychology. Obviously this, whatever its value, is not supposed to be taught in turn to children. It is part of the course of study because of our hope that when the teacher has learned something of what research workers have found out about children, she will see them in a clearer perspective than would otherwise be the case.

Nor is the situation fundamentally different when it comes to the preparation of our college teachers. No one has as yet proved, or is likely to do so, that there is any genuine relationship between earning a doctorate by writing a treatise on the sources of *Samson Agonistes* and teaching a course in Milton to juniors. Granted a reasonable amount of aesthetic intelligence, one no doubt could manage a wholly satisfactory semester with only the text and a convenient manual. The academic accessories probably do little more than befog the student's mind. But the fact that the instructor has the long trek to the doctorate behind him does enable him and his students to see each other in a wholly different and more invigorating light than either would otherwise manage, for through this companionship a young man at a desk will gain some insight into the processes of the exploration of the knowable. Fichte thus instructed the scholar: he "is to forget what he has accomplished as soon as it is accomplished, and is to think constantly of what he must still do." To have lived for some time in communion with such a scholar will be for many a young person as exhilarating, and one must immediately add as humbling, as standing on a Darien peak.

It appears unlikely that the situation is wholly different insofar as other callings are concerned. A candidate for appointment to the foreign service will have to know whatever that service at the time deems important, including how to write and speak a foreign language. But having painfully mastered Spanish, he will normally find himself in Timbuctoo or Saigon as a vice-consul, dutifully writing out visa prescriptions or practicing minor roles in the eternal drama of commerce. The average chemist will become a member of some chain gang of scientists marshaled like a posse for ferreting out a new explosive or antibiotic. And the political scientist, fresh from the study of the arcana of government, will be fortunate if he can pass a civil service examination and proceed daily to chores with the Housing Authority or the Bureau of the Budget. But if somewhere along the road such a student has caught a glimpse of the "city" as seen by a man for whom the span between Plato and Quincy Wright does not exhaust the vision of that "city" as it has been or may be, he will not sleep without dreams.

"Closeted with the Unknown"

I believe we may therefore conclude that as education proceeds it does not lose sight of the purposiveness implicit in its beginnings—namely, the imparting of knowledge—but will, when it is wisely conceived, also reckon constantly with the ultimate objective, which is sharing the life of the scholar, poet, and saint. As a matter of fact, it will be driven to do so by the passion of the best students it serves. These will question the knowledge of their teachers but never the awe of them as they stand on the brink of discovery. Thoreau in his time asked whether Concord could not "hire some Abelard to lecture us." The query seems to be universal, save possibly when men have become uninhibitedly utilitarian. It seeks wisdom for the many through the contemplation of the one. And whether the answer be given in terms of experimental science, or in those of the speculative intellect as with the Greeks, or in those of the prophecy embedded in the Hebraic tradition, or in those of mysticism, either Christian or Oriental, it will be in the final analysis the celestial fruit of a wedding between the "I" and the "Thou," to use Martin Buber's pertinent phrasing. We begin with the communication to others of the easily known in order that at long last we may find ourselves closeted with the unknown. Only if we are so placed, at least waveringly, hesitatingly, fleetingly, can we mortal beings acquire the sense of comedy and tragedy, of the holy and the profane, that gives us the stature to which it is our destiny to aspire.

Perhaps we may now venture to define the liberal arts as follows: a course of study designed to encourage tentatively integrated learning about man's most fruitful insights into himself and the reality about him, so that a student may feel the texture of the known in order to be able to realize, sooner or later, that this is only the garment of the unknown. If the known were the whole of being, we should have no answer to Newman's question about Scaliger: How could so much learning have passed through the mind of one man—and why did it pass? Aquinas in his day held that the ultimate properties of being must remain unknown, just as the potential existence of a human creature cannot cease to be enigmatical. To think of molecular movement going on constantly inside a baseball thrown to a hitter is merely to tease oneself out of thought. If the psychiatrist could map out his patient's psyche, the therapeutic task would be less impossible. He cannot do so. In the final analysis there is relatively little we can really know of other men, save that we aspire to the truth about them.

Let me add a few comments which regrettably are more addicted to the vice of generalization than could be wished. First, the educator must realize that what the storerooms of the past contain is indispensable treasure. Of necessity he will challenge the accusation of pedantry constantly leveled against him, but he cannot function unless this charge is in some measure justified. He must have books about him that few other men read. The genesis and progression of ideas he will observe with a reverence other men do not feel. Indeed, one may go so far as to say that the great teacher has a genuine affection for the past, which makes the sharp lighting up of any of its moods or features a memorable experience. But he must avoid like the plague every form of dotage leading to the assumption that he or any other human being exists for the sake of knowing what is already known.

Yesterday must be for the teaching scholar the coast line on which he can stand before plunging into the unplummeted and perilous sea of tomorrow. Therefore, secondly, education must accept as a kind of law that even the rediscovery of the past must have relevance for the present. A man will be worth his salt if he sees quite clearly that his life will be worth while only if, at some moment at least, he is visited by a creative and illuminating intuition of reality. Thus gifted, the scholar will not disparage what is called the "creative," though he will cling to his role of critic. Because he himself has passed through the open door of the mind, he will respect pioneer intellectual effort, no matter how seemingly revolutionary or unexplored the forward thrust may be.

It follows that education is in part the preservation not merely of what has been learned but also of the traditions, the methods, of learning. These are several, not singular. One individual's best way of learning may be quite different from another's. One nation, to some extent conditioned by historical environment, will not learn most adequately in the same way another nation does. We may note in passing that this is probably the principal discovery made by the Soviets in their satellite areas. Why should anybody take it for granted that all teachers must subscribe, for their souls' salvation, to a single formula? But if any teacher be a canny person, he will certainly weigh methods that have proved useful for other teachers, and he will be as objective in evaluating them as a purchasing agent is when examining samples of leather. He will consider the advice of Comenius and Kirchensteiner, Loyola and Dewey, Ulich and Livingstone.

What Is Truth?

"Truth," said Pestalozzi the optimist, indicating therewith his approach to education, "is a medicine which takes hold." What shall we say about the meaning of these terms? In other words, how shall education conceive of the real and the good?

When education declares that its function is to find and to teach truth, to what is it committing itself? The completely frank answer in the actual existing situation is: to not very much. There are vast numbers of students and teachers, at all academic levels, for whom the task assigned is merely to absorb and emit a specified quantity of information. Of course it is expected that the data imparted will be reasonably accurate from the giver's and the receiver's point of view—that the class will not say *le escargots* or assert that Shakespeare began his career by writing *The Tempest*. Therewith truth has become mere accuracy in remembering determinable data. It is of considerable interest to note that the majority of the vocal critics of our schools, like most of the schools' supporters, seem to want no more of education than this. The difference between them is merely one of emphasis on certain facts as being more valuable than others.

This version of "truth" seems inadequate to me. Nor is it satisfactory to take a similarly restricted definition of what Pestalozzi may have meant by "medicine." To assume that "good citizens" can be made to emerge from the schools as hot cross buns do from a bakery is to take a benign view indeed of human nature and the teaching profession. This happens to be what never happens. To be sure, good schools are often effective conditioning devices. If youngsters can be induced to absorb moral maxims into their blood streams at a sufficiently early age, the effect may be relatively lasting. But moral precepts will congeal as indigestible lumps unless they can be fused with the drift of the intellect and the genuine drive of the will. Character is never formed as aught save conscience, and this means a living commitment to sublimation of the self. That most reasonably gifted men and women wish to make that commitment is, I think, fortunately true. Experience seems to indicate that the family, the school, and the church can intensify this desire and direct it to good ends by awakening the child's joy and pride in belonging. A youngster who is jubilantly confident of the stature of his preparatory school will wish to be worthy of it, and this longing may endure through life. And it seems indubitable that the influence of the church is proportionate to its ability to evoke affectionate trust in its practice of the holy life. This may seem as if man were here being doomed to becoming "organization man." Aristotle long since so doomed him, as the evidence required. What alone matters is the quality of his gregariousness.

"Truth" as education must conceive of it is, then, primarily awareness of the vital activity of the receptive, creative human mind face to face with reality in the whole of its illusory overtness and its revealing concealment. It is on the one hand "man thinking," and on the other, that which can be seized and held in thought. "Truth" therefore cannot be for any wise teacher merely "what he troweth," to borrow Newman's words, because while awareness must be vividly personal, it is nevertheless bound to the whole with hoops firmer than steel. Here is a brief comment on a characteristic trait of Aquinas, taken from a recent book by Joseph Pieper, *The Silence of St. Thomas*: "The same intrepidity made him ask, in his *Commentary on the Book of Job*, whether Job's conversation with the Lord God did not violate reverence—to which he gave the almost outrageous answer that truth does not change according to the standing of the person to whom it is addressed. He who speaks truthfully is invulnerable, no matter who may be his adversary."

What is here meant by "truth" is a firm grip on some part of reality. The earth does spin round; there *was* a process of evolution, though we may never fully know how it operated. The right to discover and report such truths is the most inviolable of rights. But if a man proceeds to assert that any part of the true is the whole, if he construe his article as being the encyclopedia, he is as gravely in error as would be the planner who believed that if he built a city of skyscrapers there would be no traffic problem.

The Pressure of Technology

Accordingly, here are the poles between which education moves in practice: the scholar's free, creative, but rigorously controlled awareness of the cosmic or human verity that he holds with awe in his hands, and his humble, submissive realization that this little, precious though it be, is only like one of the diamonds on Cecil Rhodes' plain. This is why it is utterly senseless and life-destroying to hold that education can be either purely scientific or not concerned with science. In the wake of the eerie excitement caused by Russia's ability to push a satellite into outer space, we seemed for a time wholly to forget that for years education in the United States had been veering strongly to a one-sided concern with engineering and other forms of applied science, and that we were in grave danger of losing our collective dedication to the deeper forms of contemplation, whether they were concerned with mathematics or psychoanalysis, metaphysics or pure poetry.

Having been told over and over again that the United States could have sent a rocket to the moon years ago had it been so minded and willing to foot the bill, why should we now imagine that safety can be found only in thicker dabs of science on the schoolboy's bread?

The reason why the veering alluded to has taken place is of course this: The impact of scientific discovery on our modes of living is so great that we are all caught up into a Heraclitean world. Cellulose fiber makes growing cotton on sun-parched fields a dubiously profitable venture, and vegetable oils deprive the cow of a major reason for being. Indeed, one by one the animals become superfluous save when dead. Yet who can doubt that a stern appraisal of our people's ability to live in the world that is now its companion, day in and out, will reveal glaring weaknesses —widespread inability to cope with the leisure that is the by-product of technology, and a resulting softness of mind, heart, and hand; the lack of impulse to enter into the cultural worlds of other peoples, past and present; and above all a hankering after spurious kinds of "peace of mind," as if these might not prove to be the ultimate enfeebling narcotics.

If what has been said is in a measure correct, a number of conclusions are suggested, some few of which will be advanced here with the requisite intrepidity.

Singling Out the Adventurous

First, it must be obvious that education can proceed in its full glory and significance only insofar as it is concerned with those for whom it is not merely an obligation but also primarily and increasingly an adventure. Young scholars must be chosen and not simply endured. While Maritain and the Harvard Report on *General Education in a Free Society* are quite right in holding that some measure of liberal education is the privilege of all citizens, it remains as certain as anything can well be that even in the most democratic of societies many students will fail to move beyond the stage at which knowledge is only communicated fact, either because they are unable to do so or because the journey does not interest them. Those who are eager and able to embark on the *adventure* of education should be singled out as soon as possible, freed of crippling economic handicaps, and made to realize that the training of the mind is at least as rigorous as the training of the body. To continue to accept the lowest common pupil denominator as the norm is to doom the potential intellectual power of the nation to turning somersaults around the statue of Huckleberry Finn.

One happy result of emphasis on pupil selectivity would be that at long last we should be able to train teachers in a relatively rational manner. There are candidates for the profession able and willing to go with unquenchable enthusiasm to the task of guiding the unfolding creative mind. Others will be more at home with the larger numbers for whom awakened interest is the only lure. And there will be some who, sensing perhaps a vocation akin to that of nursing, will concern themselves with young people who are in a sense abnormal, because of handicaps or some lesion of the will. As things are now, in most colleges and preparatory schools the teacher has little notion of what he is expected to accomplish. He knows only that his work is with youth, and he usually finds himself in as impossible a situation as is the driver of a twenty-four-mule team some of whose charges are halt and lame while others are eager for the road. It is no wonder that problems of teacher morale exist, particularly in schools compelled to assemble in the same rooms youngsters who should no doubt be in jail and the sons and daughters of parents who have long been devotedly interested in the progress of the human intelligence.

Freedom for Teacher and Student

Some clarification of what is meant by the freedom of the teacher seems highly desirable at this point. Inviolability of the mind when it is aware of truth is the inner radiance of every free society. It is the "single string," to use Donne's phrase, that cements scholar and teacher in comradeship and mutual respect. But one cannot conclude that the same freedom should be claimed for the imparting of information, *unless this is actually the communication of truth in the sense defined.* For instance, the historian who might contend that the Roman *limes* was a deposit of quicklime should be free to say so only until somebody finds him out. Or again, a mathematician who has not in a measure kept abreast of developments in his field can hardly claim a natural right to remain in a state of ignorance. But a student of the Roman past who advances a new hypothesis concerning the meaning of *limes* must have complete freedom to publish it, no matter how startling the contentions or how inconclusive the argument may seem. Failure to make this distinction is responsible for a widespread reputable skepticism about academic freedom.

Conversely, the proper freedom of a pupil does not consist in doing what he wishes. I am persuaded that, once young people have progressed beyond the years with which Madame Montessori was concerned, they rather wistfully expect someone to tell them what to do and how. This does not mean, to be sure, that they will wear hair shirts with pleasure. But few statements can be made with greater assurance than that pupil satisfaction and response are far greater in exacting high schools such as Hunter or Brooklyn Technical than they are in makeshift mental factories for which the football season is the major academic event.

At higher levels the fledgling young scholar has every right to expect that his own dawning awareness of part of reality will be accorded mature, critical respect. Who has really learned to teach who has not at some time realized that a young mind can light up a scene which has hitherto been dark? When I dealt with a class concerned with some aspects of English verse, it was not a commentary by a distinguished critic that I used to clarify a stanza by Marvell but an essay written by a Harvard senior.

It is at this point that the marvelous utility of student discussion should be adduced. Young people do not suffer one another's foolishness gladly. Indeed, they are loath to accept the mutual exchange of wisdom. Each young person has in the company of his fellows, however, a priceless abrasiveness, an abrupt and vigorous way of proceeding from enmity to affection and back again, which is like sprouting combative antlers of the mind. How good and fruitful the college campus is (as Newman indicated a century ago) on which thought etches itself out in jagged contour during student debate! Can we not all look back gratefully and see ourselves limping by reason of the bruises earned in such struggles and the depths that had to be leaped over, but still having in the end weary but exhilarated companionship? Alas that we should lose this skill later on! I am sadly reminded that in the days of yore Henry Mencken and Stuart Sherman, the first a stout brew of Nietzsche and *Simplicissimus,* the second a glass of pure humanistic port, were wont to assail each other with uninhibited vehemence. If I had my way no student would graduate who had not had a similar glorious row at some time.

Finally there is the harassing but unavoidable ground on which "truth" and "medicine" meet. Wisdom can never be synonymous with knowledge. There is no graver peril to which modern man can be exposed than surmising that prudence is automatically built into his application of the insights that he has acquired. As he succumbs to this error, he becomes a thing that can be used rather than a man deciding of what use he can be. This we have seen with implacable clarity in the moral callousness of the gifted—scientists and engineers, jurists and writers—who have served tyrants. We shall see it even more plainly in the manipulation of minds by new and subtler forms of propaganda.

Education Incomplete without Theology

If education is to be what I have said it is, namely, "awareness of the vital activity of the receptive, creative mind face to face with reality in the whole of its illusory overtness and its revealing concealment," how can it complete its assignment unless it throws light from every available source on the questions asked by Albert Einstein about modern scientific man: "Has he not in an effort characterized by being intellectual only forgotten his responsibility and his dignity? A man who is inwardly free and loyal to his conscience can, it is true, be destroyed, but he cannot be turned into a slave or a blind tool." These things Max Horkheimer had in mind when, returning to Germany from exile to become rector of the bombed-out University of Frankfurt, he established chairs of Protestant and Catholic theology without being personally a devotee of either. He believed that some light might be cast on the queries of Einstein by a discipline that has played and still plays a mighty role in the drama of the West. Most as-surely he was not thinking of an acrimonious debating society, nor did he acquire one. The European university seems to realize that a theological faculty consists of educated men and not of self-appointed functionaries either in a kind of hypothetical Office of the Inquisition or in a club of intellectuals the primary activity of which is to blackball the parson.

It is not too much to say that the immaturity, or it might be better to say the incompleteness, of American culture manifests itself at no point so clearly as it does when religious issues are under discussion. There are many reasons why this is so, but the principal one undoubtedly is that theology has been studied so far away from the main stream of university life. Quite without knowing it, we have agreed with Tito and Rákosi in banning religion to the rectory.

The fact of the matter is that thinking in theological terms, even amidst the turmoil of recent social tragedy, has attained heights of pertinence and influence that cannot be whisked out of being. It is at least probable that the books of Père Teilhard de Chardin will outlive those of Sartre. I shall say no more than that I find myself wishing American education could face the ultimate questions concerning the nature of human existence with the same willingness to discuss the whole of the evidence that I find elsewhere in the world. To me it appears rather odd that we may all be blown to kingdom come without having allowed the American university a fair chance to talk about whether man is immortal. Not that a faculty resolution on the subject would be particularly reassuring. But at least if I were a young, inquisitive person, I should prefer to be slain on such a tremendous scene after having weighed all the evidence concerning my survival. It seems a pity to deprive the fledgling American intellectual of that opportunity.

⁂ *Discussion*

VAN DEN HAAG: President Shuster has remarked that, contrary to what he understood Professor Hook to say, humanist education is not responsible for the rise of a political system; and he suspects that limited technological education is. I want to say I don't think that it is either. I don't think that the Russian educational system was responsible for Bolshevism, any more than the German educational system for Nazism. But there is some suspicion that not the system but education *per se* may have had something to do with the rise of new and undesirable political systems. Perhaps too much education, or education of the wrong people. So, perhaps one thing that you ought to consider, in addition to considering *what* education to give, is *how much*; and whether educating too many people, and people not overly able to profit from education, may not only be a waste of resources but also have unfavorable political consequences.

HOOK: I did not mean to imply that Fascism was the result of studies in the classical and traditional curriculum; I think totalitarianism in Western Europe is a social phenomenon. I merely pointed out that upholders of traditional education in this country have insisted that the classical curriculum is essential for democratic citizenship, on the theory that, because the rulers of the past were brought up on this curriculum, now that we have democracy, in which every man is a ruler, we ought to give them the same kind of education that these past rulers had. And all I meant to say was that if we have any empirical evidence, it indicates that this kind of education did not immunize people against totalitarianism. If I had to make a bet, I would say a country that had received the kind of education we get in our country, even if its intellectual standards are not so high as in Europe, would be more immune to contagions of this sort, because of its individualism and its I'm-from-Missouri attitude.

We are committed to educating everybody. This is our really revolutionary contribution to education, begun with Jefferson and Horace Mann: anybody has the right to the best education he can get. If we accept that, we face different kinds of problems from what the Europeans face.

SHUSTER: I am glad Professor Hook has clarified his position on that point. I would say in reply to Mr. Vandenhaag that nobody is going to stop the American people from getting as much education as they want. They will even take bad education. What I tried to say is that you cannot improve the American educational system unless you improve the teaching. I think that the assumption that nearly anyone can teach, regardless of what kind of ability or dedication he has, is what drives us to the brink of ruin more than any other single phenomenon.

Is Religion a Condition of Morality?

HOOK: President Schuster raised a fundamental theoretical issue in his discussion of the place of theology in education. I would agree that no liberal-arts education can be given without a systematic and even sympathetic study of theology and philosophy. But it has to be critical. And you cannot justify religion in the curriculum on the ground of its necessity for supporting morality; man's conceptions of morality precede religion. Moreover, there is no evidence that individuals who have been brought up on a belief in theology are better people than individuals whose morality is derived from other sources. The empirical evidence shows that in communities where church-going is distributed among the population in a ratio of 75 per cent to 25 per cent, the prison population shows that 95 per cent of it are believers and 5 per cent are not.

I should concede, of course, that religion has had a great role as a support of morality, but the moral insights are always independent of the theological justifications.

SHUSTER: I think that your theology is a little primitive, since it has never been assumed in any theological system which I am familiar with that moral values are somehow separate from human experience. The faith of Thomas Aquinas was built on it.

When you tell me that in a community in which 75 per cent are Puerto Rican Catholics, to name them, 95 per cent of those in jail are from among them, I would like to point out to you that not 95 per cent of the St. Francis's of Assisi have ever been in jail, except for a good reason. So when you identify a statistic with fact, you are not being terribly empirical.

HOOK: No, no; I think you misunderstood me. I did not mean to maintain that the study of religion is a causal factor which would increase the prison population. All I meant to say was that I saw no evidence that if we were interested in moral behavior, we could get children or adolescents to behave morally by giving them religious instruction. And I think that sociologists would agree with me.

Where morality is concerned, it is the cultivation of the imagination that is important. Most of the things that juvenile delinquents do, for example, seem to be based on the complete absence of understanding of the impact of their actions upon others. A good deal of what we call immorality is thoughtlessness which results from a limited vision.

I have forgotten who said that if we were all seated at the same table no one would go hungry. It is impossible to consume food selfishly in the sight of a starving man or a starving child. In one sense we are all seated at the same table in life. And if we used intelligence and imagination to bring the faces of people who are distant closer, we might accomplish more. When I taught in elementary school, I had a tough school, with classrooms littered with all sorts of papers. One day I wrote on the board, "Somebody's mother cleans the room." This sounds corny, but from that time on that room was so spic and span that the cleaning woman complained to me that I was taking her job away. It was a little thing, but suddenly something came home to these pupils and their habits changed as the result of a widening of their imagination. If I had said to them, "People who are untidy will burn in hell," or "Cleanliness is next to godliness," it wouldn't have had any effect upon them.

KUSCH: I think the discussion here grows out of a failure to distinguish religion from theology. If you interpret theology in this literal sense—as a body of beliefs which assert, for example, that children who leave chewing gum on the floor will literally burn in hell—I agree that it has no place in the curriculum; it is the propagation of fairy tales. But in the broad sense religion represents the accumulated experience of mankind, with moral attitudes which man has taken throughout history, like loving your mother. It consists of judgments about the values, goals, and meaning of life, not all of which you can put on a rational basis.

On Trying to Teach Ideals

TUMIN: May I second Mr. Kusch's remarks and go a little further? Isn't a prior question raised here by the assumption that morals should be taught? Should we attempt explicitly to teach morals?

HOOK: Do you imply that teachers should not raise questions about slavery and desegregation?

TUMIN: I mean more than that. Is the teacher committed to taking a moral stand about them?

HOOK: That is another question. I don't say that the teacher should have a stand. But should he discuss the moral question? I think on some things the teacher should take a position; I should say he should take a position on segregation; I think such a stand is right and is not controversial. On some other things, he should content himself with exploring the issue, for example capitalism and socialism. He should present both sides, and not take a position.

TUMIN: Why? I don't understand why he should take a position on segregation but not on capitalism.

HOOK: Because there I think there are basic values as to which there is less likelihood of agreement. We are not so sure about the comparative merits of socialistic and capitalistic systems as we are that it is unfair to deprive a person of the right to life and liberty merely because of the color of his skin.

SHUSTER: All that I have advocated is that religion should be taught at the university level. I believe that religion in this country is entitled, for any number of reasons, to have at the university level a thoroughly reputable school of theology, in the same way as at the University of Tübingen and at the Sorbonne. I don't see how this could do anything but benefit us.

BUCHANAN: Perhaps this is the right place to introduce a concern which we may call the idea of "the common truth," by analogy with "the common good." The common truth is something that lies in the back of all our minds as the nature of things, or the way things are, the presuppositions that lie behind our explicit knowledge, what implicitly controls our laboratories and our public life. Part of what Mr. Shuster was saying applies to pushing the frontier of discovery toward the common truth as well as toward the individual truth. And it is our business, in Mr. Hook's phrase, to be continually critical, in the sense of trying to evoke a clearer picture of what this is.

This bears on theology in the most profound way, it seems to me. If I understand theology at all, one of its primary aims is to find out what lies in this implicit and perhaps vague but overarching vision that we all live with —in our conceptions, for example, of the solar system or the theory of evolution. Theological doctrines lie back of both of these, and not merely historically. They are very much alive today.

DIEKHOFF: I would like to make a point that carries out what I think Mr. Buchanan has been saying. A few years ago Yale established a program of religion for its students. The thing that first struck me when I saw the plan was that the Department of Religion included an anthropologist, a historian, a psychologist, but no theologian. There was no one here who regarded theology as a body of knowledge. It was something to be studied by the other disciplines. I asked a member of the committee at Yale whether theology should not be included as a discipline, and his answer was, "Yes, but what theology?" This is almost an irrelevant question. It seems to me that what the student at Yale needs is to realize that there are people concerned with this discipline as a body of knowledge. If you study it only by other disciplines as a historical discipline, a psychological one, an anthropological one, you are missing the boat.

VAN DEN HAAG: I want to bring this discussion down a little bit from the celestial heights to which it has soared to address myself to the question whether moral values can be taught. It seems that it would be desirable to teach them. The question is, can we?

It has been suggested that a teacher of arithmetic can use the subject to teach the moral value of accuracy. What precisely does that mean? Is it a moral value simply to say, "This is the efficient technique of arithmetic"? If we undertake to teach fair play to children does not this basically mean teaching that if they play fair, people will like them or they will become popular? It seems to me that when you teach the moral value of fair play, you must have in mind something valuable independently of what effects it may have on your co-players, or whether it will make you popular. If we do mean that, I wonder how we can teach it. By teaching, we mean demonstrating in some way, proving, and how could that be done? Desirable as it may be, I wonder whether moral values can be taught in any direct way.

KENNEDY: As you know, in one of Plato's dialogues the question is raised, Why can you teach people horsemanship, but cannot teach them virtue? I would say this is a misleading question, if you take it in any literal sense, because I think horsemanship is a virtue. There is such a thing as being a good horseman and being a bad horseman, being a good arithmetician and being a bad one.

What we are talking about, it seems to me, is the kind and type of habits which the individual acquires. Every one of us who is a parent teaches virtue if we inculcate habits of industry, tidiness, honesty, and accuracy in the home. Everyone knows that there is a code attached to each of the professions. While a man is studying in medical school, he doesn't get lectures on medical ethics; if he does, they are window dressing; but, working along with other people, he acquires a certain identification with them. He finds himself a member of a corps; he is imbued with a certain *esprit de corps;* he has a certain kind of morale; unconsciously he soaks in a medical ethos.

The same thing happens, in its own context, in a law school, or in a scientific institute, or in a school of theology. So it seems to me that the question we are really asking here is not how one can teach virtue in the didactic sense. The real question is whether there is some way that we can continue and prolong in the school the type of habit formation that occurs for the young child in the home—a type which occurs also in the professional schools. I don't think it is a hopeless or unrealistic question at all, though I do believe that it is much more difficult to accomplish what I am talking about in the elementary and secondary schools than in a law school, or in your own home.

BESTOR: To what degree should a state-supported day school make the pretense, or uphold the idea, of teaching moral and religious ideas, other than through the kind of indirection which Mr. Kennedy has suggested? In view of the doctrine of separation between church and state, which

I believe in, and which applies to the public schools, I would suggest that the public schools must announce more limited purposes than those which can be announced by a church school, a residential school, or any school in which church and state are related.

NAGEL: I am still very much puzzled by the recommendation that theology should be a formal part of instruction, even at the university or college level. Certainly to the extent that students become familiar with the history of philosophy, they are exposed to a variety of theologies. You read Plato, you read Augustine, you read St. Thomas, and so on down the line.

Isn't this studying theology? If something more is required, what more? Do you recommend that some one position should be taught as the truth? That, I believe, would be suicidal. I would agree with Mr. Kusch that most of these theologies, if you take them as accounts of the universe, and interpret them and reinterpret them in a non-literal way, are subtle and symbolical ways of saying something that is a common truth, common, that is, to all of them. But if you take them literally with a reasonable attempt to understand the intent of the writer, I should say they are all fairy tales. I would like to be informed what is meant when theology is recommended as part of the curriculum.

SHUSTER: What you are doing, Professor Nagel, is assuming that the essential function of the university is to provide an adequate philosophy. If we bring in some work identified as theological, you are entirely in favor of that. But if we bring in some individual who has training as a theologian, you look on him with a degree of terror modified by humor, and say, "This is an individual who does not belong in a university." Let me, therefore, go on to suggest that there are a number of people calling themselves theologians who, insofar as most of us can tell, are relatively intelligent human beings. Why not let them go ahead and see what they can do?

NAGEL: But President Shuster, I wonder if you would be specific. A number of these men already have semi-official positions at Columbia. Reinhold Niebuhr is a theologian. Paul Tillich is a theologian. Many of their views are discussed in philosophy courses, sometimes in complete agreement with the position taken by the teacher, and I think that is perfectly all right. I certainly would not object to a teacher of philosophy subscribing to Tillich's ideas.

SHUSTER: Would you object to the theologian subscribing to yours?

NAGEL: Not at all.

SHUSTER: That is all I am arguing.

NAGEL: But I don't see anything novel in the suggestion you are making.

SHUSTER: The novel suggestion is that every university should have a school of theology, as you have two at Columbia University, both very good. I think that Harvard University, which has an illustrious department of literary studies, would profit very much by adding an equally illustrious school of theology.

LINDSAY: May I ask a practical question? Granting that schools of theology are very useful in universities and colleges, do you feel that they achieve their effect simply because they are there and make their offerings to students, or do you feel that ultimately compulsory courses in these subjects should be made part of the curriculum leading to the bachelor's degree, in order that their influence should be really effective?

SHUSTER: The answer to the latter question is unqualifiedly "no."

LINDSAY: Then how are they going to have their effect?

SHUSTER: Take the Sorbonne, for example. My old and very revered professor, Bergson, has long since departed, but I think I would say that of all the men who were there in the days when I knew the university, his was the most extraordinary influence; and certainly in his later periods Bergson was more of a theologian than a philosopher. Today at the Sorbonne you have some excellent theologians. These people are respected members of the community. Their influence is felt in the same way as is that of a great philosopher, or of a great mathematician—a highly reputable intellectual influence in every sense. If I go on now again to my favorite spot on earth, which is Tübingen, I think you cannot divorce the intellectual history of Germany from the influence of the school of theology at Tübingen.

KENNEDY: I cannot see the point of disagreement. Whitehead a few years ago was at Harvard. Tillich is there now, and no doubt gets many students. Nobody I know of objects to Bergson or to these men. What are we really differing about?

HOOK: In the Washington Square College at N.Y.U. there is a Department of Religion, and courses are offered in comparative religion, Judaism, Christianity, existentialism.

We sometimes exchange teachers. But what those of us who talk as we do are afraid of is something else, the argument that religion is necessary for salvation, that you have got to teach religion, therefore, in such a way as to save students, and that you save them by giving them the truth. At the same time you try to protect the truth against critical examination. I would be prepared to examine any subject and permit any man who is qualified to teach it, so long as he is committed to the same spirit of inquiry in the study of religion that we follow in all other disciplines. But think what an outcry would arise if somebody made a critical study of Christianity in the classroom, and then announced his conviction that Christ was not only not a divine figure, but was not a historical figure!

SHUSTER: I agree with a great deal of what you say, but not, of course, with Santayana's phrase that religion is an imaginative echo of things natural, of moral and human experience, and that we can appreciate it as poetry. Any theologian I know of, regardless of what school he may belong to, would say that he makes a claim to a distinctive knowledge, of a kind which is not considered in the study of philosophy.

HOOK: You are interested not only in natural theology, but also in sacred theology.

SHUSTER: There is no question about that.

❊ *Arthur Bestor*

❊ *John L. Childs*

❊ *and Panel*

2

Education and the American Scene

ARTHUR BESTOR

"In its education," Admiral Rickover has said, paraphrasing Lord Haldane, "the soul of a people mirrors itself." A mirror, one must remember, is undiscriminating. It reflects the good and the bad, the beautiful and the ugly, with crystalline impartiality. Education is such a mirror. Defects in our educational system are reflections of weaknesses and shortcomings in our national life. This is a hard truth to accept, but a truth nonetheless. The danger lies in confusing explanation with justification. Because racial discrimination can be explained historically is no reason for viewing it in any other light than as an abomination. So with defects and weaknesses in American education. To explain them is not to condone them.

Fatalistic acceptance of weaknesses in education is equally unwarranted. The educational system is influenced, but not determined, by social forces. Of all the institutions of society (except possibly religion), education enjoys the greatest measure of autonomy. This is a precious autonomy, to be defended with unremitting tenacity, to be enlarged whenever and wherever the opportunity offers. The independence of education from social pressures must be defended not merely for the sake of education but primarily for the sake of society. Education is almost the only force *within* society that is capable, in some measure, of *altering* society. Its power in this respect can easily be exaggerated, and can even be crippled by such exaggeration. Professor George S. Counts's famous question, *Dare the School Build a New Social Order?* is as absurd as would be the question, "Dare Archimedes move the earth?" Neither is a matter of daring. The question is whether one can find a fulcrum outside society on which to rest the lever that might move society. Education has no such fulcrum. On the other hand, it is not absurd to suppose that education, working within society but retaining its power to discriminate among the purposes that society presses upon it, can bring about changes not only within the schoolroom but also, gradually and cumulatively, throughout society itself.

Education, if we would be realistic, is a minority force among the forces of society. Peculiarly applicable to education, therefore, is a pregnant sentence of Thoreau's: "A minority is powerless while it conforms to the majority; it is not even a minority then; but it is irresistible when it clogs by its whole weight." The school and the college have an obligation—*to society*—to clog by the friction of their resistance those movements in society that tend toward intellectual and cultural degradation. This requires not daring but will, a less spectacular virtue but one that is perhaps more truly heroic.

American education has been shaped to its present form by the forces in American society, forces which have been largely unresisted or even perversely reinforced. It can be reshaped by sternly resisting certain of these forces and by deliberately enhancing the strength of others. American education must be so reshaped if the United States is not to follow the path of degeneration blazed by other nations that in the past have complacently cherished their vices equally with their virtues, from a cozy feeling that both were their very own. The duty of American educators is to make the discriminations that are necessary if social forces are to be so directed as to revitalize American education, and, through education, American society.

The Three Areas of Concern

Schooling is part of the process by which infants eventually become adult members of society. It is an important part of the process, but only a part. Other agencies of society are also engaged in deliberately shaping the future of young persons. The task of the school can be defined only with reference to these other agencies, each with its particular sphere of competence and hence of responsibility.

In the last analysis, there are three areas of such great concern that every organized community provides some form of deliberate training for them. First of all, in even the most primitive society there is training for the practical tasks on which the livelihood of all depends. In the second place, every society provides elaborate means for indoctrinating its young members in the mores of the society, for transmitting its cultural traits and its ethical system. In the third place (though perhaps only in societies that we can call civilized), deliberate training is provided in the use of the intellectual tools that the civilization has developed: reading, writing, and arithmetic, at the lowest level; logic, history, mathematics, and science, or their equivalents, at higher levels. Until recent times, such training, in contradistinction to training of the first two kinds, has been provided only for a minority. Today universal literacy, the simplest index to the prevalence of at least rudimentary intellectual training, has become the accepted ideal of virtually every society and the actual achievement of many.

Each of these three kinds of training is closely related to a particular institution or group of institutions in society.

Training in practical skills is obviously related to the system of production. This functional relationship takes its most natural form in apprenticeship, a program, most literally, of "learning by doing." Apprenticeship, whether called by that name or not, can extend, and does extend, from the level of unskilled labor to the most complex professions. The physician's internship in a hospital is perhaps the best example of the latter, and is a proof that the essential principle is neither obsolete nor likely ever to become so. Where the problem is one of applying specific techniques in definite practical situations, the skill involved can be learned in no better place than on the job. To try to teach "know-how" in a schoolroom is to substitute a woefully inefficient and unrealistic procedure for an efficient and realistic one. As occupations and professions become increasingly complex, of course, more and more of the things that must be learned are transferred to the classroom. But the training that is so transferred is not—or should not be—training in "know-how," but training in the *intellectual components* of the profession, that is to say, the scholarly and scientific disciplines underlying it.

The second form of deliberate training undertaken by every society—indoctrination in the attitudes, customs, and standards of the culture—is associated, functionally as well as traditionally, with the family in the first instance, with religion in the second, with the institutions and ceremonies of the organized state in the third. This great category of training includes instruction and exhortation on such diverse matters as loyalty, the relations of the sexes, the norms of personality, morality in all its varied significances, acceptable conduct within a group and among groups, customary procedures and ceremonies, hierarchies of value—the list could be endless. Powerful group emotions are associated with all these things, powerful taboos exist, powerful forces can be unleashed against those guilty of deviation or even of criticism. Lapses in social conditioning can occur, of course, and even occasional breakdowns of social control. But it is naïve in the extreme to imagine that the agencies of society that operate in this realm are weak, helpless, or ineffectual. In fact, they marshal the most powerful forces that can operate upon the mind and spirit of man.

The third form of education is what we may call *liberal* education, or training in the scholarly and scientific disciplines basic to intellectual life. That there is such a thing as intellectual training, that it is distinguishable from job training and social conditioning—these are propositions which I find it necessary to defend in certain circles. Allow me to assume that I need not argue them here, but that I may proceed at once to examine the provision society makes for intellectual training.

The Distinctive Function of the School

The school, the college, and the university were created to perform a specific and recognized function. Their facilities and techniques—classrooms, libraries, laboratories, recitations, lectures, seminars, and examinations—were designed and developed for the particular purpose of intellectual training. To enable the school to carry out any other function, it must be altered and adapted, and its performance in the new role is usually haphazard, fumbling, and defective. Moreover, if intellectual training is pushed aside or neglected by schools and colleges, society is thereby impoverished of intellectual training, because it possesses no other resources, no other agencies, no other techniques for making up the loss. That the primary function of the educational system is to furnish intellectual training is as completely self-evident as any statement that can possibly be made about the function of a social institution, whether one approaches the matter from the point of view of logic or history or sociology.

The distribution of functions that I have described is never more than approximate, of course. Most institutions perform, in an incidental and indirect way, functions that belong primarily to institutions of another sort. Thus business and industry, not only in the "breaking in" of employees but also in advertising, attempt a good deal of social indoctrination, particularly in what are held to be the "economic virtues." The home carries on a good deal of vocational instruction and also a good deal of intellectual training. The school, even in its strictly academic work, maintains, and therefore helps to inculcate, the ethical standards of the surrounding culture, whenever issues involving these standards—for example, the matter of honesty in examinations—arise in the classroom. In providing the intellectual foundations for professional work, moreover, the school at times cannot avoid crossing the line that theoretically separates intellectual training from apprenticeship. A blurring of lines, in the degree which these examples represent, is both natural and inescapable, and it raises no question worth discussing.

Problems of tremendous magnitude, however, are created by any wholesale transfer of functions from one group of social institutions to another. The problems are particularly acute when one specialized agency suddenly projects its power over areas formerly the responsibility of other agencies. Such a situation is characteristically a revolutionary one, as when the church in effect takes over the state to create a theocracy, or when the state absorbs into itself the psychic potencies of the church to create an integral religion of nationalism, or when the business community captures the instruments of government to create an oligarchy, or when government seizes control over industry to create a communistic regime. These are extreme historical instances, no doubt, but they are suggestive. At the very least, they raise the question whether the wholesale transfer of social functions, in any given instance, is in fact necessary, justifiable, and beneficent. The mere fact that such a transfer results from—can only result from—insistent social pressures does not prove the inevitability of the development, and certainly furnishes no answer whatever to the question whether the change represents progress or degeneration.

Expansion of the Functions of the School

That American public schools have enormously expanded their functions is so obvious a fact that I do not suppose the point calls for elaboration or demonstration. Those who most vigorously oppose the point of view I take on educational policy do not deny this expansion of scope; indeed they acknowledge it and take pride in it. One of the most influential statements of the function of the public school, *Planning for American Youth,* a program published by the National Association of Secondary-School Principals, says: "Youth have specific needs they recognize; society makes certain requirements of all youth; together these form a pattern of common educational needs. . . . It is the Job of the School to Meet the Common and the Specific Individual Needs of Youth."

The "needs" that are particularized (in the ten points that make up the body of the statement) include those forms of training that I have described as job training and social conditioning. The responsibility of the school, in other words, is supposed to extend to all the areas in which society has customarily furnished some form of deliberate training.

This concept of the school as possessing a comprehensive and virtually unlimited social responsibility is itself a social concept, the product of social forces which can be traced historically. To trace them, however, is not the primary task, because explanation is not justification. There is no reason to believe that the social forces that produced this concept of the school were irresistible forces. The development of education in other countries seems to show that forces working in this particular direction can be resisted, if one wishes to resist them. And those who defend the American concept of an indefinitely extended responsibility for the school do not argue that the United States was *obliged* to accept this view of the school, but that the nation was *wise* to accept it.

Is it really necessary or desirable for the school to expand its responsibility to this extent? In fact, is the school capable of discharging such extended responsibility?

Can it perform the tasks involved without fatally neglecting its primary social obligation—that of providing intellectual training?

The questions I am asking have reference to the *curriculum* of the school—the organized course of study conducted in the classroom and directed by the teacher. Young people are under the supervision of the school for certain hours in addition to those for which classroom work is scheduled. During these hours, many kinds of activities can be carried on and, if necessary, can be required: athletics, debate, shopwork, cooking, driver training, and so on. These are extracurricular activities. They are highly desirable, and in planning them the distinctions I have made between job training, social conditioning, and intellectual training have no particular relevance. In its *extra-curricular* activities, the school may properly cater to any of the "felt needs" of young people. It may consciously undertake to foster and advance social purposes broader than those of intellectual training. Through its extracurricular activities, indeed, the school can often be extremely effective in redirecting the energies of young people, which otherwise may take an erratic course. The most traditionally minded educators have always recognized the importance of extracurricular activities in shaping the characters of young men and women. "The Battle of Waterloo," Wellington is supposed to have said, "was won on the playing fields of Eton"—the playing fields, be it noted, not the classrooms. Other kinds of battles, perhaps more important, were won in the latter. No one, I believe, has ever suggested that Eton should shut up its classrooms and turn the whole day over to sports.

The dominant educational philosophy in the United States today, however, repudiates this distinction between curricular and extracurricular activities. Its spokesmen are not saying, as I have just said, that the school can assume responsibility for vocational training and social conditioning in such hours of a student's time as the school may happen to control outside the regularly scheduled classroom periods. They are saying that the school *as part of the curriculum* should undertake to accomplish wide-ranging social objectives in addition to intellectual training. For great masses of students, this philosophy asserts, the pursuit of such other objectives may properly *replace* intellectual training.

This philosophy closes its eyes to the fact that the American public school is a *day school.* Students are under the supervision of the school for no more than half (and usually much less than half) of each waking day, for only five days out of each week's seven. During the greater part of his conscious life, a student is within the sphere of control of other institutions of society, upon which responsibility also rests. There must be a distribution of function among the agencies of society because there is a distribution of time among them. The school has responsibility for part of a child's upbringing because it controls, by law, part of his time. *Which* part of the child's upbringing, given this distribution of function, is the peculiar and inescapable responsibility of the school?

This question does not arise in a residential or boarding school, of which Eton is an example. Here the school is responsible for the entire life of the young person in its charge. The school stands *in loco parentis,* exercising the authority and assuming the responsibility of the home. It usually sponsors religious services and provides for religious guidance, thus making itself the channel through which the church performs its functions. In such a situation, the influence of every social agency except the school is suspended, and the school not only can but must assume responsibility for every variety of deliberate training that is provided. Significantly, schools of this kind invariably preserve inviolate the distinction between curricular and extracurricular activities, a distinction that is theoretically far less necessary for them than for day schools. That such schools believe no good purpose would be served by abandoning the distinction is a telling argument against permitting the distinction to be broken down in day schools, public or private.

The point is so simple that it can be stated in quantitative terms. The hours that a day-pupil spends in school are roughly equal in number to the hours that a residential school sets aside for formal classroom instruction. Extracurricular activities are scheduled by the residential school in other hours than these. A student in a public day school can expect to receive an education of equal depth and intensity only if classroom periods are preserved inviolate for serious intellectual labor and only if homework is required in an amount equivalent to that which a residential school expects from scheduled study periods. A day school that permits extracurricular activities to intrude upon the hours of classroom instruction is giving its students an inferior education.

If the day school—specifically the American public school—is to undertake vast programs of social conditioning, it should demand from society the additional allotments of student time, as well as the additional allotments of money, necessary to carry out such programs. If the school is to take over the responsibilities of the home, for instance, it must take time away from the home life of the student and must be vested with the disciplinary authority of the home. A strong argument could be made for creating in the United States a large number of full-time residential schools, in which public funds would provide both subsistence and tuition for the students enrolled. Such schools would be capable of assuming the responsibilities that American educationists are talking about. But educationists are not urging the creation of such schools. They are asking the public day schools to assume these wide-ranging responsibilities without providing the appropriate means and the indispensable time for doing so. Their proposals involve not the expansion of extracurricular activities to embrace most of the rest of a student's day and most of his evening and week-end hours, but the replacement of much of the academic curriculum by activities of another sort.

Is the All-Purpose School a Necessity?

Do the societal needs of the present day call for an indefinite extension of the functions and responsibilities of the school—an extension that would involve remaking the curriculum? One cannot answer by pointing to the many problems of society that are not being adequately met. The adequacy with which social problems are handled by any society depends largely upon the appropriateness of the agencies that are developed to meet them, the exactitude with which the means devised are adapted to the ends in view.

To take a simple illustration, the discontent engendered by maldistribution of wealth can be met in several different ways: by increasing the police force of the state to repress disorders; by altering the social order through revolution; by strengthening the otherworldly emphasis of organized religion to produce greater psychological contentment ("You'll get pie in the sky when you die," sneered the old Wobbly song); by increasing productivity through technological change; by altering the incidence of taxation; by enacting minimum-wage laws and other welfare legislation; by developing the agencies of collective bargaining, especially trade-unions. Each of these methods is obviously capable of producing some change in the given societal situation. In terms of desirability and long-run effectiveness, however, there are obviously marked differences between the various possible programs of action.

Some proposals represent the creation of appropriate agencies, some merely the seizure of power by existing but inappropriate ones. This latter point is important. A struggle for power is always involved, however concealed by professions of altruistic benevolence. Those who aspire to direct the process of social adjustment—whether government officials or members of the clergy or union leaders or school authorities—are rival parties in interest. Each is predisposed, consciously or unconsciously, to interpret the existence of social problems as a call to enlarge the functions of the institution with which he is immediately concerned. Skepticism with respect to these conflicting claims is the only safe attitude for an independent citizen.

The argument that the public day school in twentieth-century America must assume responsibility over wide-ranging areas of social concern apart from intellectual training is, at bottom, an argument that other agencies of society are not capable of, and cannot be made capable of, dealing with the problems involved. This argument I find completely unconvincing.

Job Training No Business of the Schools

So far as job training is concerned the argument verges on the preposterous. Never before has industry itself provided such comprehensive programs of on-the-job instruction at every level from the most elementary manual skills to the most complex technological and managerial ones.

The armed forces do the same. On the other hand, never before has the high school been less fitted to provide such skills. In the first place, extreme specialization has created a diversity of occupations impossible for any school curriculum to encompass. In the second place, the mechanical equipment and the technical processes connected with the modern assembly line are far too complex and costly for any school to reproduce even in miniature; hence the instruction given on the simple and outmoded equipment of even the best-equipped high-school shop is pathetically unrelated to the "real-life" situations presented by an up-to-date factory.

Roughly the same situation obtains in matters connected with the home. Manufacturers of household appliances provide elaborate instructions for their use, and some (such as sewing machine companies) offer extended courses of personal instruction. Most make some provision for installation and service. Popular magazines provide detailed recipes and instructions in the use of prepared foodstuffs and in the elegancies of home entertainment. It is true that the modern home offers less instruction than the home of a few generations ago in spinning, weaving, sewing, baking bread, laundering, and ironing. Few youngsters know how to harness a horse or milk a cow. This proves that household activities have changed, not that the home has lost the art of imparting practical skills to the young. The home has never been perfectly successful in passing on the household arts, of course; but the assumption that it has become or is becoming less effective in passing on the arts that are actually used is unsupported by any substantial evidence.

Automation in industry and the mechanization of the home have made the school an obsolescent instrument for *vocational* instruction. At the same time, the rising technological level of actual industrial operations, and even of actual household processes, has created a sharply increasing demand for precisely those *intellectual* skills that the school is particularly designed to impart: reading, mathematics, scientific understanding, and the rest. The English classroom, the chemistry laboratory, and the blackboard covered with mathematical symbols are the "functional" parts of a school today, not the woodworking shop and the homemaking room. These latter, quaint survivals from a simpler age, have considerable extracurricular value, but that they are necessary to sustain a modern technological economy is sheer fantasy.

The Perils of Social Conditioning

The wholesale extension of the public schools' responsibility into the realm of "social conditioning" is fraught with even graver dangers and is based upon equally dubious premises. Particularly dubious is the assumption that the forces that operate to produce social cohesion, conformity to the mores, and loyalty to accepted institutions have been all but fatally weakened in our society. Because of this collapse, the argument runs, the school must step into the breach by offering comprehensive programs of "life-adjustment" education.

This represents, in my judgment, a completely unrealistic appraisal of the strength of the forces that operate in this realm. Not only are they powerful to begin with, but their pressure is being constantly augmented as the new devices of mass communication are pressed into service. The possible omnipotence of the forces devoted to social conditioning ought to concern thoughtful men and women, most of all those engaged in education. In the nature of things, training in this realm is basically training in conformity. Law and order rather than devotion to principle, frictionless personal relationships rather than rationality, acceptance of group decisions rather than basic integrity of mind—these are the effects sought after (perhaps the only ones that can be sought after) by impersonal agencies of social control. No leaven of criticism lightens this heavy mass; creativity of mind is not an asset but an annoyance; dissent is not a virtue but a stumbling block.

The glowing spark of intellectual independence, which social conditioning is most apt to quench, can be kept alive in the school if the development of critical intelligence remains its overriding objective. Freedom to think—which means nothing unless it means freedom to think differently —can be society's most precious gift to itself. The first duty of a school is to defend and cherish it. This means resistance to the pressures for social and cultural conformity wherever they may arise. Church and state are less menacing to intellectual freedom in America today than are the anonymous forces in the community that insist on like-mindedness, on "belongingness" and "togetherness," on the kind of "other-direction" that Riesman has described.

Social conditioning, "life-adjustment," can only mean reinforcement of these pressures when undertaken by a public day school. Quite different is the situation in a residential school, which stands, as it were, between the student and the immediate pressures of the community. The latter can be fended off, and the school can discriminate deliberately between the ideals it will foster and those it will repudiate as unworthy. The social conditioning that such a school undertakes will be partly nonconformist; it will give its students some powers of "inner-direction" (to borrow Riesman's distinction again), some ability to resist the seductions of "other-direction." The public day school—supported by taxes derived from the community, subject to immediate community pressure, recognizably an agent of local government and of the state—lacks any such power to resist conformism, unless it takes the high ground that its proper task is to develop critical, independent, well-informed judgment by means of disciplined intellectual training. Not the indiscriminate molding of attitudes desired by the community, but the deliberate molding of a specifically *intellectual* attitude, is the function and the responsibility of a publicly supported school in a society that wishes to preserve its vital freedoms.

The adjustment to life that we must strive for through the school is the kind of adjustment that results from applying the varied resources and the developed powers of a mature and disciplined intellect to each successive problem as it arises. Adjustment in this highest sense *is* an outcome of education. It is not an outcome that can be reached by short cuts, by a miscellany of experiences, by playroom imitations of the mere externals of adult activity. "There is no 'royal road' to geometry," said Euclid to his sovereign, Ptolemy I. "There is no royal road to intelligent citizenship" is the message that educators should deliver to the sovereign people of today. Serious, sustained, systematic labor, in libraries, laboratories, and classrooms, is the only way of producing educated men and women in the twentieth century, as in every preceding century.

A Requirement of Democracy

Serious, sustained, systematic study is the one way in a democracy, as it is the one way under a monarchy or an aristocracy. The crucial difference is that in a genuine democracy the school system provides this kind of intellectual training for all its citizens instead of for a selected few. If a nation permits something else to be substituted for intellectual training in the upbringing of substantial groups of its citizens, then it is no more democratic in its educational system than any other country that relegates part of its population to inferior status by furnishing that part with inferior education. Intellectual training for some of the people, vocational training and life-adjustment for the rest, is the epitome of a class-structured educational philosophy. Those who preach it in the United States are simply repeating the arguments that have been used by opponents of democratic education in every age and in every country. These opponents have always asserted, as frightening numbers of American educationists today assert, that the majority of men either have no need for serious intellectual training or are incapable of receiving it, and that for them mere training for the job or for adjustment to life will suffice.

How has it happened that a democratic society like the United States, committed to the democratization of education, has produced in the twentieth century an educational theory that is patently antidemocratic in its fundamental assumptions and its practical proposals? This is one of those historical developments, like the emergence of dictatorship out of a basically egalitarian revolution, that appear at first glance paradoxical but are nevertheless susceptible of historical explanation.

A diffused anti-intellectualism is characteristic of any group and any society that is largely uneducated. Anti-intellectualism is a social force to be reckoned with in any historical period; it is a force of critical importance in any period of educational expansion or democratization. How it is dealt with by those responsible for education determines the success or the failure of their efforts to develop an effective school system.

The anti-intellectualism of which I speak is not necessarily a virulent hatred of intellectual values; it is more often a mere indifference to them. It ordinarily does not imply opposition to schooling as such; rather it involves contempt for the outcome of schooling. Though jealousy and class antagonism have something to do with anti-intellectual attitudes, the fundamental cause is ignorance itself. Uneducated men and women not only do not know, they do not know what it is they do not know. The outward results of possessing education, measurable in terms of wealth and prestige, are observed by them, but not its inner characteristics. In this view from the bottom, intellectual training often appears to be simply a particular form of job training, that is to say, training for well-paid and respectable jobs. It may even appear to be training in an arbitrary set of subjects to which completely artificial distinctions have been attached—a kind of protective tariff imposed by the privileged classes to keep out possible competitors.

The problem of extending education to groups and classes that have hitherto been deprived of it is, therefore, a twofold problem. It is not enough simply to provide schools. It is essential also to make perfectly clear the precise purpose that schools are to serve. If fundamental purposes are not defined and adhered to, anti-intellectualism may rush in to fill the vacuum, with the result that schools may become allies of the very thing that they were designed to eradicate, namely ignorance.

Where Educational Responsibility Lies

The responsibility for defining the primary purpose of any institution is a professional responsibility devolving upon those who already understand the matters with which the institution is to deal. Similar is the responsibility of the medical profession to define the purpose of a hospital, prescribing what it should do and what it must never consent to do, no matter what public pressure is put upon it. This is in no sense an attempt to dictate to the public; the latter can decide whether they wish to support schools or hospitals or not. The question is whether a professional man may permit the perversion to different purposes of an institution he is charged with defending. He has a professional obligation to combat every such destructive trend and to resign if he is defeated.

The central problem is to make use of social forces in such a way as to build up an institution, instead of permitting such forces to tear it down. Take, for example, the powerful motive of economic striving. Properly utilized, it can contribute immensely to the raising of the school's standards of intellectual competence. Improperly utilized, it can so warp the school program in the direction of narrow vocationalism as to destroy its integrity and all but destroy its usefulness.

Horace Mann, to take one notable educational statesman, recognized the distinction. He appealed frankly to economic motives in rallying support for public-school improvement, but he did not surrender the intellectual purpose of the school in so doing. In his Twelfth Annual Report he wrote: "For the creation of wealth, then—for the existence of a wealthy people and a wealthy nation—intelligence is the grand condition. The number of improvers will increase, as the intellectual constituency, if I may so call it, increases. . . . The greater the proportion of minds in any community which are educated, and the more thorough and complete the education which is given them, the more rapidly . . . will that community advance in all the means of enjoyment and elevation."

Until the beginning of the twentieth century and somewhat beyond, social forces in the United States were effectively harnessed to the task of educational expansion and improvement because educational leaders translated the demands of society into terms appropriate to the school. Educational advancement ceased when educationists deserted the ideal of disciplined intelligence and accepted the fallacious notion that the character and content of school programs should be determined by the "felt needs"—that is, the immediate, uncriticized, short-sighted demands—of those in society who had least comprehension of what education was for and how it accomplished its ends. No wonder health departed, once medicines began to be prescribed not by the physician out of his knowledge but by the patient out of his ignorance.

Educational statesmanship consists in directing the insistent demands of society into such channels that the intellectual needs of society, felt or unfelt, will be effectively met. By adhering firmly to the distinctions that educated men know to exist, by uniting the learned world in defense of standards, by giving academic recognition only to those who demonstrate intellectual achievement, educators can use the vague and heedless desires of the ignorant themselves as the motive power of a system that will dispel ignorance. The secret is to love the sinner but not his sin.

JOHN L. CHILDS

I shall begin with three assumptions. First, organized education is the kind of enterprise in which factors of time, place, and culture are centrally important. The pertinence of this assumption has been freshly shown by the volume of criticism to which American education has been subjected since Soviet Russia pioneered in getting earth satellites into orbit. Since cultural evaluations are inescapable in all educational construction, these evaluations should be made openly, with responsibility for what is chosen and rejected.

Secondly, we have entered a period when world affairs should play a central part in the development of American educational thought and practice. I am in accord with those who hold that we are involved in a grim struggle in which the stakes are nothing less than the preservation of the values of democratic civilization, and that since this global struggle may last for decades, the kind of education our young get in the public schools may have decisive bearing on the ultimate outcome. Good as our educational program has been, and recognizing that many of its present functions must be continued, it is nevertheless unsound to assume that it can be "education as usual" for a generation which *must* serve literally as the trustees of the heritage of Western civilization.

Thirdly, I assume that foreign and domestic affairs are so interrelated that strength in the world sphere requires strength in the home sphere. The present world struggle is most fundamentally a struggle between divergent patterns of human civilization. Our cause will succeed or fail at this elemental level of civilizational character and response.

Our country is involved in the kind of struggle in which we must possess the industrial and military power required to discourage Communist aggression. Yet we can achieve our supreme ends only as we are able to maintain a world situation in which it will never become necessary to make use of ultimate means of destruction. In thus emphasizing the cultural and political aspect of the task, I do not imply at all that the territorial, the technological, the scientific, and the military aspects of the struggle are not important. Strength in these dimensions of the task is obviously of the first order of importance, and we should not let it be sapped by wishful yearnings for peace that are not robust enough to confront stubborn realities. I do suggest, however, that these modes of carrying on the struggle can contribute to the attainment of our basic ends only as they are governed by political, economic, cultural, and moral policies that have the tendency to strengthen the forces of freedom in all parts of the world—Communist as well as non-Communist.

It is from this perspective that I have chosen four major educational problems for consideration.

1. The Future of the Common School

Today there is wide recognition of the fact that the public school must enjoy more adequate financial support. Students of taxation also consider it unlikely that this additional revenue can be secured by higher rates in the general property tax, the present main source of local school funds. This means that federal aid to education is needed today, not only to equalize educational opportunity among the various regions and states, but also to provide the funds that are required to improve instruction in all parts of the nation. Thus far all attempts to pass a federal aid bill for the general support of education have failed. This failure is due to a number of factors, but the most important one is the inability of the supporters of organized education in our country to agree on the major purposes and provisions of a federal aid bill. These differences are deep because they involve the future of the American common school.

As Lawrence A. Cremin emphasizes in his historical study, *The American Common School,* this school was originally conceived as an institution that would serve American democracy by providing the young of all groups of our society an opportunity to live, study, and play together and thereby develop those appreciations, understandings, and allegiances that are the ultimate basis of our lives.

The record shows that in state after state the response to the plan of a common school was immediate and positive. By 1850, Cremin reports, of the 3,644,928 enrolled in all the educational institutions of the country, no less than 3,354,011, or a little more than 90 per cent, were enrolled in some kind of public facility.

On the other hand, the American common school has never functioned as a universal school. Various religious groups have continued to maintain schools supported by private funds. Certain favored economic groups all along have preferred to send their children to private schools. In the South and in some communities of the North, colored children have been deprived, by inherited patterns of discrimination and segregation, of the opportunity to attend the common school. Some staunch supporters of the principle of the common school believe that American education is strengthened, not weakened, by the acceptance of a sphere for private initiative in matters educational.

In this connection, however, one point should be clearly stated. Our common school system cannot serve its historic function under conditions in which a nationally organized and numerically significant group repudiates the whole conception of a common school, and moves with resolution to develop a comprehensive, alternative educational system of its own. Public education in the United States is now confronted with precisely this critical condition, and until it is resolved the educational forces of our country will remain divided. The Roman Catholic Church has developed a church school system, designed not to supplement and improve the program of the common school, but rather to supplant it so far as the education of its own children is concerned. About two-thirds of all

Roman Catholic children are currently attending parochial elementary schools, and about two-fifths of the available children are enrolled in parochial high schools.

Roman Catholic educational leaders have also endorsed bills that have contained provisions of direct tax support for their schools. Indeed, they believe they are so clearly entitled to this support that they condemn and oppose any federal aid measure that does not provide substantial subsidies for their own educational program.

There is a strong possibility that cases involving these complex issues of religious liberty, of church and state, and of the American plan of a common public school, will soon be taken to our courts, and eventually to the United States Supreme Court. Decisions from the Supreme Court on marginal issues in this field reveal confusion and conflict in the views of the justices. It is highly desirable that mature analysis of these fundamental problems be made before the lawyers develop their briefs and the judges begin to pronounce on the issues involved.

2. *National Unity and the Segregated School*

Thus the historic principle of a common school in our country has been weakened by the refusal of the prelates of the Roman Catholic Church to accept this kind of public school as the agency for the education of their children. This same separatist tendency is also exhibited by certain Protestant leaders. For example, Henry P. Van Dusen, president of Union Theological Seminary, has recently declared: "Unless religious instruction can be included in the program of the public school, Protestant church leaders will be driven increasingly to the expedient of the church-sponsored school." In the case of all these church-related schools—Roman Catholic, Protestant, and Jewish—the segregation from the public and common school has been self-imposed; it has not resulted from the refusal of the public school to keep its doors open to all, irrespective of creed or religious affiliation.

The situation is radically different when we consider the pattern of segregation that colored children have suffered under the educational arrangements of many of our states. In these states the public school has never been a common school, for it has been organized into two branches, and the colored children have been barred from attending the school in which the white children receive their education. For a considerable period we assumed that the demands of equality could be met by this segregated school system, provided that the same quality of education was given in each division—a condition which was never approximated, although in recent years real gains have been achieved by the colored schools.

The heart of the 1954 decision of the Supreme Court was its unqualified endorsement of the principle of the common school. In its opinion, segregation had been tried and it had failed to meet the standard of justice in the sphere of education. A unanimous Supreme Court, which included justices reared in the South as well as in the North, declared that "in the field of public education the doctrine of 'separate but equal' has no place. Separate educational facilities are inherently unequal." In a later ruling the Court decreed that the measures required to organize a common school open to all, irrespective of race or color, should be taken with all "deliberate speed." These were words of power from our highest Court. They were literally heard around the world—in the Communist countries, in the uncommitted nations and areas, and in the liberal-democratic nations. The decision strengthened the democratic cause in its moral foundations.

But that was several years ago. Unfortunately, the

record of deeds is not nearly so eloquent as the legal formulation. In the last analysis, however, it will be the deeds and not the words that will determine whether this decision to eradicate historic patterns of segregation and discrimination is to count as solid achievement for the democratic cause.

We should not minimize the important steps that have already been taken. In a number of border areas our educational leaders have shown vision and courage, and important beginnings have been made in the process of transforming the segregated school into an integrated school. Two great failures, however, mar the record we have made thus far. Because of these failures it has been relatively simple for the extreme white racists to take the initiative.

One of these was the failure of President Eisenhower to use the great powers of his office to develop a positive public opinion on the basic issues. Much as we should commend President Eisenhower for taking the bold step of using the forces of public authority to uphold the law, we should also remember, as Talleyrand once observed, that "we can do everything with bayonets except sit on them." In other words, the cooperation of the liberal forces of the South will be required to develop a nonsegregated school system, and our President did little to create a situation in which these progressive forces could undertake the difficult things that must be done if we are to attain our social end.

Equally serious has been the failure of organized educational groups to take a resolute part in the development of the means by which the cause of the integrated school could be advanced. These educational organizations are national in membership, and they include colored as well as white teachers. After all, the problem of developing a common school is not primarily a military or a legal problem; it is a problem in the field of organized education, and it involves the art of human relationships, an art in which the educator is supposed to have a special competence. Great credit is due individual educators for the able leadership they have given in difficult situations, but the profession of education as an organized movement has not measured up to its responsibilities. Up to the present, religious leaders, lawyers, and publicists have done more in this field than have educators as an organized professional body. The failure of the President to do more to create a public opinion on this fundamental question must be charged, at least in part, to the weak initiative the educational profession has given.

Although the focus of this segregation problem is in the South, it is nevertheless a national problem. This is so for two reasons. In the first place, attitudes of race prejudice are present in all parts of the country, and they tend to weaken vigorous response to the lead that the Supreme Court has given. In the second place, the moral strength of our nation in world affairs is very much involved in what we do in the sphere of race relations. Our nation cannot inspire confidence abroad in its democratic purposes unless it seeks to establish the principle of equality of opportunity in its social and educational practices at home.

3. Education for the Atomic Age

One of the revolutionary changes in American civilization is the dominant role government has come to play. A new mentality is beginning to emerge, which is adapted to this change in our patterns of living. Leaders of the two major parties, for example, in their discussion of the kind of measures that should be taken to overcome the recent recession, have tended to assume that the economic welfare of the country is a direct responsibility of government. As a matter of fact the huge defense budget makes the government the outstanding factor in the national economy. In addition, the social security program, the agricultural subsidies, the highway, housing, and other public works programs, the acceptance of full employment as an official goal of government, and regulative fiscal and tax policies, have all combined to make government a primary and pervasive influence in the lives of all citizens.

The importance of foreign affairs also has grown steadily. And in this field governmental functions and responsibilities are apt to continue to expand until world security arrangements are achieved in which all can have confidence. Obviously, this new situation has profound implications for American education. Strong government and the preservation of the values of democracy are compatible provided we have a citizenry equipped in historical perspective and in understanding of contemporary affairs, and disciplined in methods of critical thought that qualify it to make mature judgments about public leaders and public policies. The supreme business of public education at this time is to provide our young with the kind of experiences that will equip them for their roles as citizens of a country whose power is such that it must play a crucial part in the leadership of the world's democratic forces.

This means that an anarchic pattern of individualism must pass from our schools, as it is beginning to pass from our economy and from the world operations of the nation. The welfare and the growth of the individual will remain a central objective of organized education, but that welfare and growth must be interpreted in the context of the interdependence that characterizes our domestic life and economy and that likewise characterizes the relations our country now sustains with the rest of the world. Today, the education of American children should be deliberately planned with reference to stubborn realities in the world situation and to the role our nation is obliged to take in that situation.

As many have emphasized, this involves systematic education in the disciplines of mathematics and the natural sciences. It also means that those who demonstrate special gifts and interests in these fields should be encouraged to advance as their abilities permit. Beyond all question, our technological civilization in both its domestic and world operations must have an adequate and continuing supply of people trained in the natural sciences and the various branches of engineering.

It does not follow, however, that this competence will be best developed by an authoritarian and dogmatic program in which learning from lectures and texts is substituted for first-hand experience in the development and the testing of meanings through direct experimental operations. Present critics of the new education have ignored its most basic value. John Dewey was not opposed to learning from books that recorded the discoveries already made. Nor was he in any way indifferent to the development of disciplined habits of thought in the young. He gave the title *How We Think* to one of his most important books on education. He also elaborated the principle of the activity curriculum precisely because he was deeply interested in the nurture of disciplined and resourceful minds. He believed that the young would best develop their intellectual powers if they learned to think operationally. I know of no evidence that shows that he was in error on this crucial point.

Moreover, in the determination of the present educational task it is important that our analysis of the present civilizational need be an accurate and balanced one. In order to measure up to contemporary demands our country must maintain its industrial and military strength, and this clearly means that our schools should do their part to equip young people for research in the natural sciences as well as for the technological development of whatever scientific discoveries are made. But the scientific and technological aspect is only one side of the present problem, and it is a mistake to assume that it is more fundamental and urgent than are other aspects, such as the political, the economic, the ideological, and the cultural. Indeed, the leaders of the nations associated with us in the total global program are inclined to think that we exhibit greater strength at the present time in the technological and the military phases of the problem than we exhibit in its political, cultural, and ideological phases. Our ultimate goals are economic, political, and cultural in nature. Competence in science and technology must be matched by competence in cultural understanding, in the field of human relations, and in political affairs.

In the present divided world, the youth of our country need adequate knowledge of the ideology, the purposes, the institutions, and the strengths and weaknesses of the Communist totalitarian systems. They also need education in the cultural backgrounds, the interests, and the national aspirations of the peoples of Asia and Africa, who are no longer content to have their countries serve as mere sources of raw materials for the industrial West nor merely as markets for its manufactured articles. More than ever before, our young need knowledge of the roots of our culture in Western civilization, and grasp of the historic struggles through which our democratic institutions and values were developed. Education in these historical and cultural affairs is not a luxury nor a mere adornment; it is a life necessity, for what our nation's leaders can accomplish will be determined by what the understanding of the American people enables them to do.

The achievement of this type of cultural and political maturity is no simple enterprise for a continental nation whose population has long been habituated to its own self-sufficiency, and which tends to think of education as preparation for personal and pecuniary success. The development of a curriculum adequate to accomplish these cultural objectives, and the enlisting and the educating of a teaching force equipped to nurture the young in this altered orientation, defines a civilizing task at least as important as the defense programs centering in the Pentagon.

4. The Education of American Teachers

Viewed from the standpoint of the property required to house it, the annual appropriations required to maintain it, and the numbers professionally engaged in it, organized education now ranks as one of our foremost institutions. In 1957 about thirty-eight million children were enrolled in the public and private elementary and secondary schools of our country.

By their support of these schools, the American people have shown they are aware that the scientific knowledge and the technical skills and techniques involved in the operation of an industrial society are not automatically transmitted from one generation to another.

As a people, we also have been rediscovering what the founding fathers stressed; namely, that a democratic society is a distinctive kind of society and its system of self-government presupposes an educated citizenry. The cultural perspectives and the social and political insights required to appreciate democratic values and the nature of emerging problems are no more automatically transmitted than are the technical abilities required to maintain our industrial civilization. Along with the home and the church, the school has the function of cultivating the emotional and intellectual dispositions in the young, which are the ultimate foundation of our democratic way of life.

Education today obviously must be continued throughout life. Fortunately, we have in the mass media of communication potent means for that never-ending enlightenment, providing we have the will to demand that these agencies serve educational as well as commercial and entertainment purposes. Special adult educational classes have their place, but the character of the intellectual interests Americans exhibit will be determined more largely by the daily fare contributed by radio, television, and newspapers.

It is because organized education has all these functions that many now share the view of Horace Mann and John Dewey that the educational process itself calls for serious intellectual study. Indeed, it has been increasingly recognized that teachers, school supervisors, and administrators should receive special education for their public responsibilities. It is from these convictions that normal schools, educational departments, and schools and colleges of education have developed.

These departments and schools of education have four main areas of specialized study. The first is concerned with the scientific study of human beings and human behavior, the process of human growth and development, the nature of learning, and the ways in which habits, skills, meanings, knowledge, and attitudes are acquired, together with the measurement of human abilities, individual differences, and specific educational achievements. The second is concerned with the study of American civilization in its historical and world setting from the standpoint of the introduction of the young to its ways of life, thought, and value, together with a study of present trends, problems, and potentialities. This area is also concerned with the history of organized education and with the study of contemporary systems of education as they now function in other countries. The third area is concerned with the study of the organization of the curriculum for different age groups, with due regard for what is known about the learning and maturation process, and the history, the traditions, the institutions, the problems, and the tasks of American civilization. The fourth is concerned with the management of education as a public institution, its problems of support, its relation to its community, its special services such as health and recreational activities, and, most important of all, the maintenance of conditions that will attract able teachers and enable them to carry on their work effectively.

Today these departments and schools of education are being subjected to many criticisms. A number of foundations have subsidized educational projects which have deliberately ignored these teacher-training institutions and have sought to make organized education a direct responsibility of the liberal-arts colleges. That there is substance to some of these criticisms no one should deny. Actually, some of them have been made by members of the staffs of various teachers colleges for several decades. Undoubtedly there should be closer coordination between the work of the liberal-arts colleges and the departments and schools of education in the education of teachers. But it seems to me that, just as religion, art, law, medicine, journalism, engineering, and business have their special schools and colleges, so education must continue to have its distinctive professional schools. If we are concerned to strengthen public education in our country, we shall not seek to supplant these professional educational institutions but to strengthen and improve them for their tasks.

❋ *Discussion*

CHILDS (in comment on Bestor's paper): I do not share Mr. Bestor's view that there is an inherent opposition between the intellectual and the practical—between the cultural and the vocational. The conception of an experience curriculum was developed by John Dewey not because he was indifferent to intellectual values, but rather because he was supremely concerned about them. Dewey developed his educational theory as an alternative to a formal school in which pupils were given to listening, to memorizing, to reciting. He perceived the danger of verbalism in a school program which was exclusively centered in a symbolic representation of life, and which provided little opportunity for the young to have first-hand experience of their environments—physical and social.

In his *How We Think,* Dewey emphasized that the heart of learning was learning to think. He believed that the life of the mind is related to all the activities which men have evolved in the pursuit of their interests. He perceived that it was in this matrix of primary experience—an experience of doing and undergoing, of being and having, of suffering and enjoying—that thought finds its source and its stimulus, as well as its purpose and its test. He sought to create a school environment which would provide opportunity for the young to have vital experiences

—an environment which would be composed of more than blackboards, readers, spellers, recitations, and examinations—because he was concerned to cultivate the intellectual powers of each child. Thinking, as he viewed it, involves dealing with problematical situations, the development of ideas for the resolution of these situations, and the testing of these ideas by the consequences they produce when put to the test of action.

Undoubtedly some have carried certain of his educational ideas to extremes. But all along there have been many educationists who have brought these extreme practices of their colleagues under criticism and educational correction. Dewey's educational principles do not require that we drive curricular "subjects" out of the school, or exhibit a disrespect for knowledge and rigorous intellectual work.

It seems to me that organized education in all its branches—the humanities and the sciences, the cultural and the vocational programs, the liberal-arts colleges and the professional schools, the colleges and universities and the lower schools—will better serve our country if educators seek the ways of cooperation, including that form of cooperation which is criticism, than if we divide into warring sects, each determined to destroy the other.

TUMIN: What I should like to ask Mr. Bestor is this: How much weight are you willing to give to the desires and wishes, informed or otherwise, of the vast majority of the parents of our children? And how far do you think we ought to make decisions independently of them, attempting to mold them to our purposes?

BESTOR: I think we must reshape the parents' thinking to a far greater extent than we have in the past generation.

The remaking of the curriculum in terms of uninformed, widely dispersed, and inconsistent pressures has gone much too far, and we should reverse this trend substantially.

MRS. ROBERTSON: How would you start working on the parents, who are, after all, taxpayers and voters?

BESTOR: We have the devices already. Those who are actively concerned with education and enjoy the respect of the public could undertake to define what the purposes of education are, and to embody these in examinations and requirements for diplomas or certificates.

HEINDEL: Some of Professor Bestor's comments apply to our universities as well as to our secondary schools. Has not the university been proliferating courses? There are things not unlike driver training in many of our college curricula. I have been distressed by the amount of vocational spirit in all our higher institutions, whether colleges or professional schools. We should not dodge the fact that there is a tremendous amount of vocational attitude even in the liberal arts. In one college of arts and sciences in this country there are more than a thousand courses available to undergraduates. This is a land of abundance, but is this sort of abundance helpful? Is it necessary? Is it the true application of our educational purpose?

It seems clear that we have overextended vocational and professional training in the schools. For example, we have almost professionalized nursing out of existence. Fortunately, owing to great social pressure, many people are beginning to understand that here is a noble occupation, and that maybe a great deal can be accomplished in two years instead of four years. What is happening in the profession of pharmacy? Four years used to seem enough;

now everybody is pushing for five years. Medicine is another example. Heaven knows we want our doctors and physicians to know everything they can. But we have reached a point now where a man is thirty-three years old, or perhaps thirty-four or thirty-five, before he is allowed out on his own.

My point then is this: should we not put our own university houses in order first? Where else will the intellectual attitude of this country develop, if it is not in the universities? It surely isn't there now, in my humble opinion. The mess in the Ph.D. situation is a standing disgrace to American education. To be sure, a reform movement has been started. But I think we must start at the top and try to ensure that the teachers have an intellectual motivation when they leave college. Perhaps some of this will rub off on their students.

BESTOR: I agree entirely with Dr. Heindel on all this. The proliferation of courses touches on a defect that runs from the top to the bottom of our educational system—the substitution of course credits for demonstrated knowledge. I would strongly stress the point I made about examinations. These need not be thought of in a narrow sense; they may take the form of essays, papers submitted, and so on. But giving a college degree on the basis of an accumulation of courses, considering a man trained as a historian because he had so many courses of this kind, instead of asking him to demonstrate at the end of four years that he is a historian by a wide-ranging comprehensive examination, is a fundamental fallacy.

I think, myself, that the proliferation of courses would be most effectively checked by the requirement that the student should show what he knows. At the moment, courses proliferate according to a kind of Gresham's law:

the poor drive out the good, because a student can get just as much credit for enrolling in something which isn't worth anything. I am suggesting that, instead of this, you superimpose, as in the English or continental systems, a comprehensive examination, in which whatever course a student takes counts for him only if, in fact, it can be shown to have advanced his knowledge.

Once you put in the comprehensive examination, the courses which did not contribute to knowledge would wither on the vine. In this case, then, the development of a course becomes a serious enterprise; it is more than the project of a particular man.

BUSH: I would like to mention the appalling extent of illiteracy. Every university and college has to have an army of young instructors trying to teach freshmen the rudiments of expression—a vain effort, as a rule. I think there is no excuse for having to do that sort of thing in college. It is tremendously expensive; it absorbs an enormous amount of energy; and, of course, it contributes, among other things, to converting the first two years of college into high-school work, or, for that matter, work which should have been largely accomplished in elementary school.

HOOK: In many things that have been said, it seems to be assumed that there has been a decline in the intellectual quality of the colleges and universities. Here, again, if you will excuse me, we need scientific findings; it is a problem of fact.

When I went to college, I heard my teachers talk about freshmen in the same way as Professor Bush; freshmen always have been unable to write. The atmosphere in the colleges wasn't intellectual. There was a great emphasis on football; people would say that students were more interested in the mystery of the forward pass than in politics. This was true even of Harvard. I have been told by some of my friends at Harvard, who have been teaching for a long time, that student interest in intellectual affairs is higher today than it has ever been in their experience. I think the same is true at Dartmouth, at Amherst, and at Columbia.

Dr. Bestor said, if I understand him, that our colleges and universities are really not interested in ideas and in the intellectual life; he gave the impression that everything has been crowded out by a concern for the professions. Although we can certainly improve, the fact is, unless I am mistaken, that we have made considerable progress in introducing into the colleges an intellectual content that was not there forty or fifty years ago.

BUSH: There are several other reasons that contribute to this increasing seriousness. For one thing, when I began teaching in the 1920s, it was a boom period, and youth had an easy road to wealth. Since then there has been a depression and a world war. Students in general are more serious, and the colleges in general have more careful methods of selection. There are a number of contributing elements.

A National System of Education?

BUCHANAN: Mr. Childs spoke of segregation in the schools. In addition to the segregation that we know about in the South, he suggested that private schools are in some sense segregated schools. It seems to me that the ability grouping of students for different kinds of instruction is another kind of segregation.

Now I believe it is very important that students of different abilities should be up against each other. I had the advantage, or the misfortune, whichever it may be, of having been educated up to high-school level in a district school in Vermont. There were four rooms in this school, so it wasn't really the little red schoolhouse; but the teaching was of that kind. There were at least three grades in each room. I used to listen to the people ahead of me and the people behind me. It was like the experience children have in listening to adults talk about things they don't understand. This did extraordinary things to me. I have always been grateful for it. The mutual spontaneous teaching that happens in a classroom where you have different kinds of people of different abilities is very important. Every student is himself a teacher, and this teaching of one by others is a much more important process than the activity of a teacher standing before a class and giving out this and that.

Now I jump to a dangerous idea. It seems to me we ought to be thinking about setting up a national system of education, with a national university at the top providing leadership and setting standards which would run down to the lowest level of education. It would provide for adult education of a systematic and institutionalized kind.

I remember the first time I proposed this thought: it was to R. M. MacIver when he was professor of sociology at Toronto; I think the date was 1919. And did I get a dressing-down! This was the dangerous Napoleonic system of education in France! Mr. MacIver and I have had a running fight ever since on this point.

No doubt it would frighten the American public, and it *should* frighten us, for there are dangers in it—nothing like the dangers in our present system, but real ones nevertheless. But the fright itself might do something to our educational drift; it might make us get together and really think about what the common and fundamental education should be.

We have an obligation to propose some sort of plan

as a guild of educators and teachers. We are not professionals unless we do this. If we are held back by doubts whether we can persuade the public, or how we can get taxes for the sort of thing that is necessary, we are not doing our duty. Our duty, as teachers, is to say what this country needs in education, even if it should cost a hundred times what it costs us now.

ALLEN: I share to some extent the views Mr. Buchanan has just expressed. It seems to me that the time has come when we need to re-evaluate the principle of local autonomy in school government. We have had no national system of education. We have believed that we could preserve freedom better if we were loosely organized, as far as education was concerned. But the question now arises whether it is in the national interest to maintain a school system in which schools may be as poor as they wish.

But of course if some agency or institution at a higher level is to tell the parents what schools locally should do, we have a very serious question involving our whole democratic procedure. I would like to hear more discussion as to how a uniform control could be exercised and still preserve what we cherish so greatly—freedom of choice.

VAN DEN HAAG: Mr. Buchanan's method does not function badly in Europe. But the practical question is whether it can be transferred. We must ask ourselves whether the centralization that Mr. Buchanan advocates, given the traditions that we have here, would not, instead of leading to higher standards, lead to more degenerate standards. Perhaps we should start with something such as Mr. Bestor proposed; namely, setting up national standards according to which certain degrees, for example, would be nationally recognized. That would not involve setting up a national system; how people were to achieve these standards would be left to local initiative.

BUCHANAN: I don't want to be misunderstood about impositions from above. A minimum would be worked out, and more or less required of all schools from top to bottom, but they could add anything they wanted to the minimum program. There would also be—though I was frightened when I first thought of this—inspections. State departments of education already exercise the right to inspect all schools in the state; isn't that true?

ALLEN: Yes. We don't like to use the word "inspection."

BUCHANAN: I know you don't, but I want to use the bad word, because it refers to what seems to be an unavoidable part of the system. I am thinking of what happened to the Webbs in London in connection with their work on religious and secular education. I believe it was Sidney Webb who invented a policy which has been in force now for fifty years or so. All the schools in the London County school system are required to give a certain kind of formal liberal-arts training. This is inspected by a central group. The school is at liberty to do anything else it wants to, but the fundamental arts standard has to be held to. This is simple enough, and I don't suppose we have any real doubts about it. I don't think anyone here doubts that we ought to have a minimal amount of mathematics, a minimal amount of language, in the schools. This is the kind of thing that we are agreed on before we think about it; and when we think about it, we can't dodge it. It is necessary.

ALLEN: Much could be done if we strengthened our state educational systems. I would be a little reluctant to go all the way to the federal government. It seems to me if we concentrated on strengthening the state systems of education, providing a system of inspection, if you wish to call it that, and fixing standards there, we could do a great deal of what you are talking about. At present, you have in many states two or three different educational systems. You have a state board for vocational education. You have a separate state board for general education. You have still another board for higher education. If you are going to have effective over-all planning and maintain standards all the way through, you need a single organization with strength to do the job.

BESTOR: I would differ somewhat from Mr. Buchanan in his estimate of the effectiveness of a central setting of standards and measuring of achievements. It is not inappropriate here to note the operation of English universities, which control the examinations but leave each member college a law unto itself in terms of the staff it appoints and the way it conducts instruction. Since the university can state what the examinations must be on, and what must be covered by them, you can standardize the subject matter in the sense of saying that certain things must be done. But you avoid a Napoleonic centralization and control.

It seems to me it would be in keeping with our tradition of local initiative to have the state—not a national agency—establish standards for, say, the high-school diploma. This would preserve local interest and support for the school system, without granting to the community the power to make its schools as bad as it wishes.

It would also make possible integration into a common school system of the parochial and private schools. I am not, like Mr. Childs, opposed to these on principle. It seems to me that a diversity of forums of instruction is a good thing, provided the community can be assured that the basic things, the sciences and humanities, are taught. Public examiners could examine all candidates for the diploma equally, whether they have studied in public school, private school, or parochial school, whether they have been tutored or studied on their own. I would extend this to the colleges as well. In a country with more than a thousand different colleges, the notion that the standards of the college can be set by itself, given the range from Harvard, Yale, and Princeton down to institutions that shall be nameless, seems to me impossible.

The same problem has now arisen about the Ph.D. Our assumption that the Ph.D. is at the same level in all institutions is contrary to fact.

I would like to see our system reconstructed with this examination as its central feature, rather than the establishment of a national university, or another set of schools, such as Admiral Rickover has recently advocated. I should like to see the insistence on certain standards maintained not through machine tests but by a really searching kind of examination which would provide a basis for the awarding of all academic recognition from the earliest stages to the highest. I think this would save the community responsibility that Mr. Childs eloquently spoke of and at the same time provide national standards.

How High Are Present Standards?

NAGEL: It is my impression that during the past twenty years we have managed to raise the quality of college and university instruction in this country so that the standards of our Ph.D. and lower degrees are comparable with those of European universities. I also have the impression that among the better universities of the United States the standards are fairly uniform—for instance, that the degrees in physics from Columbia, Harvard, California, Minnesota, are pretty much of the same quality. Or isn't this so? I put these comments as questions.

BESTOR: Are you talking about Ph.D.'s?

NAGEL: About both Ph.D.'s and bachelors' degrees. While I am about it, may I ask a further question? Should we not seek to set standards for degrees on an international level?

BUCHANAN: I doubt that our standards are actually high; I have the feeling that the standards are very low. But I am open to conviction. I don't know; I don't believe anyone knows. This is a dangerous position for us to be in—not knowing whether we are doing well or not.

The other point I would like to make, which bears on the international suggestion, grows out of a year I spent at Fisk University. The southern problem is terrific. I don't believe the idea of state authority is going to help us there at all.

LINDSAY: With regard to the Ph.D. generally, I don't think we have reliable facts. But I think that in some subject matters there is no question about the adequacy of our performance. The unfortunate thing, of course, is that the Ph.D. is given in a wide variety of fields. I don't want to tread on any toes, but many people feel that many of the Ph.D.'s in education, for example, won't stand comparison with the Ph.D.'s in mathematics and physics. In general, however, I don't think we need be ashamed of the standards we are maintaining for the Ph.D. in many fields in a large percentage of the graduate schools of this country.

However, I should like to raise another question.

What about the students in our colleges who are not going on to graduate work, who will never become eggheads in any sense of the word? What kind of education are we to give them? I agree with Mr. Hook that in many of the colleges the intellectual atmosphere is very good. But the difficulty I have found in my own teaching is that though many of these students who come to us from the present high-school system are full of enthusiasm for learning, they can't learn because they don't have the tools. What are you to do with them? You have to do what Professor Bush indicated we are doing; you have to provide remedial courses. It seems to me fantastic—a caricature of college education—to have to teach people how to read and write in college. It has now come to the point where this has to be done in graduate school, because the colleges don't want to do it either.

KENNEDY: I should like to get back to the question of the efficient determination of standards. Haven't we had some revealing examples of how state control may work—for instance, in the certification of teachers? The policies adopted by state legislatures in prescribing for the certification of teachers have frequently been more harmful than wise. Why should it be supposed that state legislatures, or the federal Congress, would institute a system of education better than the one we now have?

I would myself look with great apprehension upon a national system of examinations. It could become a deadly thing, a kind of Chinese system in which students were oriented toward using all the tricks they could to beat their fellows in these examinations. I can see this ruining teachers, too, because the teacher might have to sacrifice individuality and any special skill in molding a class to get the kids through certain things on time.

MRS. ROBERTSON: I would like to ask Mr. Allen a question about the Regents' examination. I am a graduate of a New York City public high school and proud to be one. Do you feel that because of the Regents' examination New York State high schools have been leveled up, and are better than high schools in neighboring states?

ALLEN: No, I don't know that that is the case. I will say this, that the good teacher does not level down; the poor teacher does. If you did not have the examination, the poor teacher would level down below it. So the good thing about the examination system is that it does assure the people of the state that a minimum prescribed program will be provided for everyone. Then it becomes our job to undertake the business of encouraging and training the teachers, of developing opportunities for going beyond these minima, and of using the system properly.

MORGENTHAU: It seems to me the diversity of standards in American education reflects a diversity in American society which cannot be overcome by any central direction of the educational system. Take, for instance, the educational system of a city like Chicago. Here you have central direction, but you have the most extreme diversity in actual performance within the system, which reflects the enormous diversity of Chicago society. You have excellent schools in excellent neighborhoods, and unspeakable schools in unspeakable neighborhoods. And I wonder how centralization on a national scale could overcome such extremes in the standards of American society.

As for the relative quality of American higher education, I have lectured and taught in several non-American countries, and I am absolutely confident in saying that the

American intellectual performance at the highest level is better than anywhere else in the world. Take France, take Switzerland, Germany, Austria, Great Britain, Japan; in none of these countries do the universities equal the vitality and creativeness of American scholarship and higher education at its best.

It is quite a different matter when it comes to such things as how to spell and how to write one's mother tongue. In these areas no doubt many other countries rank much higher than the United States.

A Commissioner Cites Cases

BOEHM: I cannot acquiesce in all of Mr. Bestor's criticisms of our schools or in the rigid school program that he seems to be advocating. A case or two will indicate the sort of monstrosities that can occur in the inflexible public school structure to which we are now bound. We had, in the largest school in our state, a student who was permitted to take two mathematics units. When he passed in the upper 25 per cent, the school authorities gravely sat down to debate which course he should get credit for! In another case, a city school system asked the Ford Foundation to come in and determine whether, if a course was given by television and the work was covered in four weeks, the students should be required to sit under someone in an auditorium to fill out the remaining two weeks of attendance called for by the Carnegie unit system, which prescribes six weeks for the course. We also had a girl who studied German in the summer, passed the examination for three years of German, and had the temerity to ask the principal whether she could take fourth-year German. The principal answered that she could, but she would not get credit for second- and third-year German—which she needed to enter medical school later—because she had not met the time-attendance requirements.

To put an end to this sort of thing, we intimated to the high-school principals of our state a few years ago that the Carnegie time units were no longer in force in Pennsylvania.

Many of the ills of our secondary schools of which Dr. Bestor complains stem directly from our colleges. He pointed out that the time spent on football often interferes with secondary-school work. There are schools which report that, out of 180 days, they get a net of 142 days of scholastic achievement, because the last period of the afternoon is given to football. But it was Princeton and Rutgers that gave us the great contribution of football. Dr. Bestor might have mentioned baton twirling. This too we inherited from the colleges. It is well to remember also that 60 per cent of all the teachers in this state come from the liberal-arts colleges and universities. If they are poor in English, some responsibility belongs with the same universities that taught us football, baton twirling, and so on.

As for the curriculum, there is a dubious cliché abroad in the United States that what is hard is necessarily good and what is easy is softening. The truth is that what is easy for one individual may be very hard for another—and very worth while for both. Music is hard for me; history is something I love, and I find it easy. So when I recently read in a survey that the students would first kick out history and, second, languages, I was distressed. What has been happening? I talked with a judge in my own county, who said that the teachers of history kill the interest in it. Perhaps it is the method that makes things hard or easy, so that we can't say of subjects themselves that what is hard is good and what is easy is enervating. I am told that in our oldest and largest medical school, anatomy is taught in four months. If a student sits up until one o'clock at night and memorizes the bones, is this good because it is hard? Not necessarily.

A word about vocational courses in our schools. Let me say that instruction in typewriting may contribute to many ends other than vocational. Are we to bar it from general education? In spite of all the vocational courses in our schools, Dr. Bestor, their graduates do qualify for admission to arts colleges; it need not necessarily be assumed that such courses seriously interfere with intellectual education.

Those of us directly concerned cannot accept the dichotomy that education must be *either* this or that. Of course our first consideration has to do with the fundamentals taught in the school, and these might be better. But the extracurricular activities that go on between four and six o'clock in the afternoon do not necessarily subtract from education in the regular school session.

I do think that our able students, and even average students, could cover in eleven years what they now do in twelve. And some of our liberal-arts colleges might go on a twelve-month basis, enabling doctors to start their careers two years sooner than now.

BESTOR: I think I agree with almost everything Dr. Boehm has been saying. I certainly feel that the measurement of achievement in time units is absurd in every way. What I am urging is more significant measurement of students' achievements than we now have.

HOOK: Mr. Bestor maintains that the multiple-track idea for education of our youngsters in the high schools is undemocratic. What we have found is that a considerable portion—10 per cent, 20 per cent, 25 per cent—of those who do not go on to college cannot master certain subjects essential for college. Do you think it undemocratic, Mr. Bestor, to give these students special instruction?

BESTOR: No one in his right mind would suggest that a student be put into a course in calculus if he hasn't reached the stage at which he can make sense of it. But this is not to say that the boy is completely incompetent to take mathematics and therefore we will give him cooking. A good education is still a good education, in terms of its essential content. I don't believe that everybody should study American constitutional history, but I believe everyone should study American history. If a boy doesn't get history, it is no substitute to make him take shop. This doesn't mean that everyone must take the same course in history, but he should not be allowed to replace it with something quite different in its effect on the way he thinks.

HOOK: I would agree that that is true for history; I would agree that it is true for English. But can you give the lowest 25 per cent of high-school students foreign languages or mathematics, beyond perhaps a little algebra? Such questions don't imply that the only alternative is to give them cooking. Give them more English, more history; give them something that will enable them to achieve greater growth, particularly if you have to go on educating them until they leave school.

My conception of vocational education is not job training. I would say of the 25 per cent who cannot acquire math or foreign languages: give them English and history, and then introduce them to certain kinds of vocational education. While they are studying machines, or the use of certain tools, you can give them work in citizenship, certain things that involve life situations.

BESTOR: I am unaware of any controlled experiment showing that 25 per cent of the population cannot under any circumstances learn a foreign language. I think you are generalizing from the I.Q. distribution and from experience that in existing set-ups with the kind of courses actually given, a certain proportion of students do in fact fail to learn. But I think one of the most important things which educational psychology could undertake is a really controlled set of experiments to find how much truth there is in the assertion that under no circumstances can persons of a particular group acquire a particular kind of knowledge.

As to your second point, we ought to distinguish between what is intellectual training and what is not. What I object to is the system of interchangeable parts by which anything studied is to be regarded as the equivalent of anything else in academic recognition. I think the curriculum should have a limited set of functions, a limited group of disciplines, and that purely academic measurements should be based upon achievements in those particular disciplines, leaving plenty of time for the other things to be done, but refusing to accept a purely vocational course as the equivalent of one involving intellectual disciplines.

Teacher Training

WOODRING: Personally, I think that the preparation of the teacher is far more important than the curriculum. I would much rather have a good teacher with a bad curriculum than a bad teacher with a good curriculum. If you had a

Socrates in every school, I wouldn't worry about curriculum—I'd just let him teach.

The pattern of teacher preparation in the United States is not well understood. We often hear the charge that all they teach in teachers colleges is methods. This, of course, is ridiculous. There is no teachers college, and never has been, that taught nothing but methods.

The standard pattern is a four-year course, starting with two years of liberal (though sometimes not sufficiently liberal) arts. The second half of the course consists of a mixture of the academic discipline to be taught with professional preparation. In western states about a fourth of the four years of work is usually given to professional preparation; in New York, the requirement, I believe, is eighteen hours. This is not all methods, by any means. It includes a unit of study which is essentially child psychology, and another which is a historical and philosophical introduction to the American school system.

The Fund for the Advancement of Education has supported projects which attempt to shift this program in the direction of more liberal studies, assuming that two years of liberal preparation is not enough, even if, on top of this, you get a major in a subject. The Fund has encouraged a program of four years of liberal education.

One of the things we quickly found is that requiring a liberal-arts degree is very different from requiring a liberal education, and that it is necessary to liberalize the pattern in the liberal-arts college as much as in the teachers college. For this reason, much of our work has been in the college of liberal arts.

We have also supported projects which include a fifth year consisting of basic professional subjects, such as child psychology, and introduction to the philosophy of education. These courses may be taken in connection with a year's internship, during which the teacher gets close supervision. This pattern has been severely attacked by professional educators, as you know.

If high-school history teachers do not know history, the proper people to accept responsibility are the people who teach history in our colleges and universities. Only about 20 per cent of our high-school history teachers are coming out of teachers colleges.

TUMIN: Where is the best state system for recruiting good teachers, and why is it best?

WOODRING: I don't know of any state system that I should want to take over and endorse wholeheartedly. And there may be wide differences within a single state.

In New York the high-school teachers come largely from the private colleges, and some from Albany State Teachers College. Albany has an excellent liberal-arts pattern which compares pretty well with those in liberal-arts colleges, and without an excessive amount of professional work. Most of the other state teachers colleges in New York—Mr. Allen can correct me if I am wrong—prepare elementary teachers, though each one also specializes in special teachers for high school—in physical education, in art or music, in home economics, and so on.

In the State of Washington, where I was for many years, there are three teachers colleges, all of which train teachers for subjects at all levels. The state has only a single, general certificate, which qualifies its holder to teach Sanskrit or any other subject, so far as the law is concerned. This seems utterly indefensible. It is based on the assumption that a teacher, like a physician, should first of all be a teacher, and can then go on to specialize later. I am not happy about the way this is working, but some people in Washington are.

I think that our whole pattern of teacher education is due for a shake-up, and is in fact undergoing one.

NAGEL: Couldn't you give us some of your own suggestions?

WOODRING: I think the big trouble in the professional education of teachers is that we have too many separate little courses. And many of the courses overlap badly. The situation is worst in the large universities, where there may be almost no communication among the instructors of the various courses. At N.Y.U. you have, I believe, more different courses in professional education than anywhere else, and no adequate communication among the people teaching it.

My solution would be to reorganize the units into two or three major ones. The first would present everything that we think important and adequately known about the nature of the learning process. The second would give the teacher an understanding of the meaning and purpose of education—the kind of thing sometimes called the "social and philosophical foundations of education."

These two units I would put in the undergraduate program, and have them count toward a liberal-arts degree. Then I would devote a fifth year to the study of methods and materials of instruction, combined with internship in a public school under an amount of supervision which would be geared to the student's initiative, resourcefulness, and ability; if he were a very able teacher, I would let him pretty much alone. I would also have weekly seminars dealing with such questions as: How do you set up and organize a course in history? What books do you use, what materials, what maps?

DIEKHOFF: It seems to me that our liberal-arts faculties have, on the whole, resigned the obligation to train teachers and left it to the schools and departments of education, about which they then complain. If we of the liberal-arts faculty spent four hours a week in high schools for a year, so that we could talk in less complete ignorance of what the problem of the secondary-school teacher is, the college would perhaps begin to take responsibility for better teaching of subject matter to the potential teachers.

WOODRING: I think you can get away from this buck-passing in only one way, and that is to make one group of people responsible for liberal-arts education at all levels. It has long been popular with college professors to blame it all on the high-school teacher, who blames it on the elementary-school teacher, who blames it on the parents, who blame it on the ancestry on the other side.

I assume that liberal education begins as soon as the child has learned to read. It goes on in an informal sense all your life. My suggestion is: let a group of people take responsibility for the total liberal education of youngsters from the age of ten or twelve to about twenty. In that eight years you may teach them history, science—anything you think is a part of a liberal program. At the end of that time test them, and if you find the program hasn't worked, go back and change the pattern until it does work. As long as we have the college, high-school, and elementary-school programs planned by separate groups of people, we shall have buck-passing.

BUTTS: I think I would differ with Mr. Woodring and with some of the proposals in the Ford approach as to when the professional education should be started. I think it would be better to take advantage of the vocational interests of students and start their professional education for teaching in the junior or senior year of college, and continue their liberal education, along with the professional, up to the master's degree. I think there is more chance of improving both liberal and professional education if both go along together in this way.

As for the teachers colleges, I don't think we need worry very much about them, because they are passing out of existence rather rapidly anyway. The general tendency is for them to become state colleges, and in some quarters they then become state universities, especially in the West. This may or may not be a good thing, but I think it is a fact.

TUMIN: Mr. Butts, assuming it were possible politically, financially, and otherwise, would you make teacher training as long as medical or legal training; that is, have a full liberal-arts course and then three or four years of professional training?

BUTTS: No, I wouldn't make it that long, by any means.

TUMIN: Assuming you had unlimited resources, would you take this to be desirable?

BUTTS: I don't believe it would be necessary, except possibly for leadership positions or for the more specialized type of training. But at the high-school level, much of the content of the teaching is in the familiar fields of the liberal arts. I think the main place where I would differ with Dr.

Woodring is that he would put much more of the responsibility for the training of teachers upon the academic courses. I would seriously differ from that, on the grounds that education is a profession, and that it needs, therefore, a special group responsible for the continuing study of education at the school level, and its whole orientation to the culture. I don't think that academic departments of liberal-arts colleges either want to, will, or can do this. So I think there should be a group of professional educators. I wish we could put to rest the mocking term "educationist." We do not speak about "medicinist" or "legalist" or members of the other honored professions in derogatory terms.

In Defense of Common Sense

MORGENTHAU: I wonder whether, as a layman and parent, I may contribute some remarks to this discussion which the professional educators may well regard as irrelevant.

We are all agreed, I think, that there is something wrong with American education on the elementary and secondary level, and we are searching for remedies for those ills. A remedy that has not been mentioned, because it lies outside the strictly academic concerns, is the introduction of a measure of common sense into the educational system, and that brings me to some autobiographical remarks.

I have two children, a boy of fifteen and a girl of twelve. The boy goes to a University of Chicago school; he went there from nursery school onward. The girl goes to another private school, the Faulkner School. When my boy was in the first grade, I asked him, the second day, "How do you like school?" He said, "I don't like it." I asked him why. And his answer was, "They don't know how to teach."

This—in a sense—mature remark has been borne out by his career in that school. For this boy, who is now a sophomore in high school and fairly intelligent, does not know how to spell. He has no idea of English grammar; he has no idea of punctuation; he does not know when to make a capital letter and when to make a small letter. But he knows a great deal about the Indian caste system, the problem of free will, and whether there are historical laws.

TUMIN: He can learn how to use commas later on.

MORGENTHAU: He will never learn it. His idea is: What difference does it make—only the content counts.

The girl, on the other hand, hasn't the slightest idea of free will, and doesn't care. She knows exactly how to spell; she seems to know everything about English grammar, punctuation, and so forth. Sometimes when she sees her brother's English composition she laughs her head off about his mistakes in spelling. The common-sense question arises, how did this happen? The girl isn't more intelligent than the boy; quite the contrary. The common-sense answer is that in the "progressive" school my son attended it was regarded as beneath the intellectual dignity of a teacher to teach such mundane things as spelling and to have children memorize mechanically certain things which can be learned only by memorization. My daughter, on the other hand, attends a school where she has to do exercises in spelling, in grammar, in synonyms, and so forth, and by a mechanical process learns what needs to be learned. The boy was never subjected to such training, and for this reason never learned. I wonder whether the high-level discussion we have had here on teacher training does not overlook this basic common-sense problem of how to teach things which can be taught only in a mechanical way.

❋ *Reinhold Niebuhr*

❋ *Hans J. Morgenthau*

❋ *and Panel*

3

Education and the World Scene

REINHOLD NIEBUHR

The world situation in which the younger generation of Americans today must live their lives and exercise the responsibilities of citizenship has three novel aspects. The task of education is to help the coming generation to measure the significance of these three aspects and to adopt creative and responsible attitudes, rather than irresponsible evasions into the asylums of hysteria or complacency.

The most obvious new aspect is the unprecedented growth of American power and responsibility in the community of nations. We emerged from the Second World War incomparably the strongest nation in the Western community. We have adjusted ourselves to this new situation of power and responsibility with some degree of grace and wisdom; but many of the problems that the new generation must face are derived from the fact that it is difficult for a nation to adjust itself to the precarious eminence of world hegemony with only such scant apprenticeship as we have had.

The second aspect of the world scene to which we must adjust ourselves is the contest with a Communist bloc of nations which, under Russian leadership, has managed to build an empire of tremendous strength. The failure of the Western nations to solve the problems of early industrialism gave an original plausibility to the Communist creed. The healthy portions of Western society have long become immune to the virus of communism. But the creed, designed for the capitalistic world and fated to defeat there, has achieved a surprising plausibility in the dark continents of Asia and Africa, where new nations are suffering from residual or still potent resentments against the imperialistic impact of Western civilization in the nineteenth century, and where pastoral and agrarian economies, immersed in poverty, desire the technical skills and the capital equipment necessary for their entrance into a technical competence and prosperity. We are faced with a formidable foe who possesses many advantages over us in appealing to the loyalty of the uncommitted nations.

The third new aspect of the world scene is the nuclear stalemate, arising from the fact that both of the great powers have enough nuclear weapons to destroy the foe, but no defense against being destroyed in the process. We are destined to wage a battle of competitive coexistence on the edge of an abyss of total war. The possibility of stumbling into this fearful war through miscalculation becomes greater and greater as guided missiles of continental and intercontinental proportions are perfected.

The Responsibilities of Power

The task of preparing the younger generation to live meaningfully and responsibly in such a world, with so many unprecedented factors of promise and peril in it, is obviously as great as any that has been faced by educators in any era. Our nation must train young people to represent it and to become agents of its power in every field of technical, political, and humanitarian service in which this power comes in contact with the world. We have not had the traditions of the colonial service, which prompted young Britons to prepare themselves for "foreign service" and to act as proconsuls in its imperium. It is surprising that we have done as well as we have with our proconsuls, recruited from the army and business and the academic life. But we need not only amateurs but professionals in this task of representing the nation. We need young people who have studied the history, the literature, and the political institutions of Asia, Africa, and the Middle East. A special word must be devoted to the importance of language studies. Our high schools, compared with European high schools, lycées, and gymnasia, have always been defective in language studies. Our universities have required two foreign languages as prerequisites for advanced degrees; but so defective have been our language studies on the high-school and university level that few doctoral candidates have ever mastered the foreign languages with sufficient competence to use them as tools of learning in their respective fields.

We are told that young Russians, next to their devotion to the natural sciences, are assiduous in learning foreign languages to prepare for contacts with the nations in their far-flung imperium.

The problem of training the specialists needed to represent us in foreign lands is, of course, only a small part of the task facing us. We must prepare the whole new generation to assess the hazards, possibilities, and perplexities that our nation faces. Foreign policy has been defined as the Achilles heel of democracy, because its problems do not impinge directly on the lives of the people, whose voice must be final in the decisions of foreign as well as of domestic policy. A high degree of trained intelligence therefore is required to make a democratic nation equal to its foreign tasks. Many of our traditional approaches to foreign policy must be challenged, and lessons will have to be unlearned before the nation can truly find itself in its unaccustomed role.

Understanding the Russians

The Russians claim to be anti-imperialistic by grace of a creed which equates capitalism with imperialism and makes any "socialist" nation anti-imperialist by definition. But in fact the Communist faith has become the basis of a cohesive international imperialism, ruled by an oligarchy which goes under the name of "democratic centralism."

The claims and pretensions of this oligarchy toward the people of the world are equally plausible and implausible. Among the former subject nations, Communism encourages nationalism and seeks to support the "victims" of European imperialism. Communism has pictured the impact of strong Western cultures upon traditional cultures of Asia and the primitive cultures of Africa as purely exploitative, disregarding the creative contributions of Western culture which made nationhood possible for the new nations (e.g., the fact that Great Britain gave India its only common language, the English language).

In Europe, on the other hand, Communism emphasizes its internationalism and defines all feelings for national identity as "national chauvinism." It holds its satellites under subjection by the power of the Red Army, and, as recent events in Hungary prove, it is quite capable of the most ruthless action for the sake of maintaining its strategic position. In the European culture, which has achieved a tolerable solution of the social problem of an industrial civilization, Communism is no great threat, particularly because its utopian internationalism and imperialism are incompatible with the highly developed national communities. But in Asia and Africa this combination of internationalism and nationalism and this promise of offering the keys to the entrance of a technical civilization are very attractive and plausible. There is no question that it will have the advantage over us for a long time to come. We must reckon with an ideological system which has assumed greater and more plausible proportions than anything known in history since the decay of the three medieval empires, the Eastern Christian, the Western Christian, and the Islamic empires.

We cannot hope to change either the Communist ideology or its structure of power by understanding these things. But we must understand them if we are not to make the vulgar mistake of believing that we are confronting merely an ephemeral phenomenon, or that Communism and Nazism are identical forms of totalitarianism. The despotism we face is much more dangerous than all these superficial theories assume. It would be a mistake to rest our analysis upon historical analogies and overlook the unique features of the Russian despotism—a despotism which is sufficiently democratic to give educational opportunities to the children of the peasants and which requires technical skill for the achievement of its goals. The success of the Russians in the conquest of outer space proves that despotism is not incompatible with scientific achievement. But we do not yet know whether the development of scientific specialists will provide the leaven of freedom for this political structure. The ideological and philosophical problems must be analyzed afresh in our educational enterprise, if we are not to flounder in a sea of contemporary events without knowing the currents and the winds that determine the momentary oscillations and the deeper currents of history underneath the momentary tumults and configurations.

Our Anti-Imperialist Empire

If it is important to understand the foe and competitor and to know the sources of his power and prestige, it is even more important to understand ourselves—not only as a nation but also as part of a European civilization and of the so-called "free world." It is a difficult task to understand ourselves in the world scene, because our situation contradicts our theories almost as much as is the case with Russia. We too are an empire and, in a different sense, anti-imperialist. The Russians are, despite their anti-imperialist philosophy, the only imperialists of modern history in the classical mold. But we must govern an empire while we are pathetically anxious not to be imperialists. Our anti-imperialism does not prevent us from exerting our power where our interests are clearly threatened, nor does it prevent such manifestations as our economic dominance in the "banana republics" of Latin America. Our anti-imperialist tradition—which was overtly violated only once, and then briefly, in the Spanish-American War and its brief aftermath in Cuba and the Philippines—is derived from the European anti-imperialism of the liberal democratic movement. But it was expressed by us more consistently than by any European nation, partly because of the circumstance of our birth in a war with an imperial power, and partly because the wide expanse of our continent made imperial expansion beyond our borders unnecessary. We therefore tend to regard the nations of Europe, though they are our allies in a common struggle, as tainted with "imperialism." Obviously we are wanting in empirical exactness, because among these allies is on the one hand Great Britain, which has successfully transmuted an empire into a commonwealth of nations, one of the greatest achievements of our day; and on the other hand there is France, which has tried to combine the old imperialist tradition with the liberal tradition of the French Revolution in incorporating Algeria into metropolitan France. Constitutionally the French maintain that "Algeria does not exist," but politically it is difficult for France either to give Algeria her independence or to grant her a rightful place in the political organization of the nation.

We still must come to terms with the imperial dimensions of our own power. We confront a superpower that is really an empire, and we are a superpower desperately anxious not to be an empire. This frequently means that we withhold the exercise of our power and responsibility in favor of the principle of collective security, supposedly embodied in the United Nations and previously in the League of Nations. From Woodrow Wilson to Dwight Eisenhower, we have expressed a touching devotion to the principle of collective security and loyalty to the instruments of that idea. Although we are now thoroughly committed to the responsibilities of power, the theories of Eisenhower are remarkably similar to those of Wilson. We exercise our responsibilities by an almost frantic loyalty to the United Nations, forgetting that loyalty to that organization is no substitute for making policy in it.

Thus in the Suez crisis we joined the Russians in the United Nations resolutions condemning the action of the British and French and ordering them to withdraw "forthwith." The assumption was that the "aggressor" nations would cease and desist out of respect for the majesty of international law. As a matter of fact, they had to bow to our power, which was greater than theirs. On 20 February, 1957, President Eisenhower, in a speech in which he justified the pressure upon Israel to leave Egyptian soil, declared: "If the United Nations once admits that international disputes can be settled by the use of force, then we will have destroyed the very foundation of the organization and our best hope of establishing a real world order." This was at a time when the Russians were in the process of brutally suppressing the Hungarian revolution. The President was not unmindful of the stark contradiction between the Russian defiance of a "real world order" and our pressure upon the Israelis to disavow the use of force when it was a matter of survival for the little nation. But he remarked that "we are entitled to expect" better conduct from the small nation because her people, like those of the United States, "are imbued with a strong religious faith and a sense of moral values."

Moralism and the Realities of Power

This curiously moralistic approach to the complex problems of the world order, which in each instance obscures not only the power realities in general but our power and responsibility in particular, is typical of American policy. It is an exaggerated version of the liberal democratic approach, which regards the power realities as not nice or as irrelevant and is particularly embarrassed by the fact that we are a great power and must exert that power. Yet we cannot refrain from exerting it, both for our sake and for the sake of the whole community of nations that depends upon our power.

The educational enterprise for this nation, in short, must include a thorough re-examination of the problems of political morality, which will help the new generation to understand that consideration of power need not be a cynical defiance of the moral order but can well be what responsible statesmanship has always been: an effort to coerce competitive and contradictory human aspirations and interests into some kind of tolerable order and justice. Such a task is a highly moral one.

This represents a tremendous task in re-education of ourselves and our nation. It is a necessary task, for without it our nation is exposed to the hazard of having power and responsibility of great importance without a scheme of values which might morally validate what we have to do for the sake of the peace of the world and the survival of our civilization. The Communist ideology gives the Communist bloc leave to exercise its powers in the name of a pseudo-universal utopian community. Our power is immediately more embarrassed because our ideological scheme recognizes only the "equal sovereignty" of all nations on the one hand and the vaguely defined community of mankind on the other hand.

The Atomic Stalemate

The final and most tragic dimension of the world scene to which we must adjust ourselves consists in the atomic stalemate in which the two great powers and the entire world are involved. We are fated to contest with Communism for the loyalty of peoples and to dispute with its imperialism every position of power in the world. But both sides are prevented from starting a decisive conflict, because the nuclear weapons spell destruction for both victors and vanquished. Thus we are fated to walk, or to wrestle, on the edge of a terrible abyss. Bertrand Russell and others think that capitulation to Communism would be preferable to the bearing of such a risk; but during the First World War he believed that capitulation to Germany would have been preferable to war. Bertrand Russell is a great philosopher; but this irrelevant advice proves, as Sidney Hook observed, that mathematical logic is not a substitute for common sense. Even a stupid statesman has the common sense to know that nations do not accept the known yoke of slavery for the sake of avoiding an unknown fate, however terrible in its known dimensions. There are others who counsel us to prepare the younger generation for coming to terms with the Russians in some disarmament agreement; and others who flee to utopia and insist that only world government will save us. Any realistic analysis of historical possibilities must come to the conclusion that disarmament agreements are the fruit, not the presupposition, of a relaxation of tensions. As for world government, it is a part of the universalism that we have previously defined as one of the characteristics of the liberal democratic inheritance. No, there are no simple ways out of this terrible dilemma. Therefore the only possibility for education is to prepare the new generation to live with the dilemma without alternate moods of hysteria and evasion.

The educational process can guide the new generation in coming to terms with this dilemma by the same tools with which it can help it to understand the other dimension of the world scene. It can delve into history to make the novel situation a little less novel. No generation has ever before faced the exact dimension of the nuclear stalemate. But many generations have faced the frustration of not being able to eliminate an evil that they abhorred; and they have finally come to realize that a common fate united the two foes—the fate of a common predicament. At the end of the religious wars in Europe, both Protestants and Catholics finally recognized that neither was able to destroy the other or push it out of Europe. Both in the end adjusted themselves to an uneasy coexistence on the same continent. Perhaps this is an inexact analogy for what our generation must learn.

It is difficult for our nation to learn, for we have never known permanent frustrations. In the days of our youth and seeming impotence we usually had our way as a nation in the end; and now in the day of our seeming omnipotence we cannot have our way. That, again, is not a new experience in history. It is only a new experience for us. Therefore the task of the educational process is to enlarge our limited experience in ways such that this frustration will become more comprehensible and bearable to us as a nation.

History is replete not only with frustrations but with unpredictable resolutions of these frustrations, and we must be prepared for these also. They prevent the frustration from weighing too heavily upon us. If we should escape the fate of being plunged into the abyss of a nuclear war, there may be possibilities of a more peaceful coexistence with the Communist power. We do not know whether or in what respect the Communist oligarchy will be leavened. The recent triumphs of Khrushchev suggest that a despotic system still requires a single despot at the apex of the pyramid of power and that not very much has changed since the death of Stalin, except that the aspiring despot is too old to reach Stalin's absolute power. On the other hand, some perceptible gains must be recorded. The power rivals, except for Beria, whom the whole oligarchy feared, have not been liquidated but given insignificant posts. Their milder treatment probably, though not necessarily, is due not only to Khrushchev's greater security but to the collective agreement of the Presidium, the power in the Communist pyramid. We do not know whether this means that the Presidium has become a genuine instrument of collective leadership in the manner in which the Whig aristocracy established the collective authority of Parliament against the monarch in England, or whether this is just another instance of the practice observed by the Roman emperors of the Augustan principate, who allowed the old senatorial aristocracy to preserve the symbols of authority while they manipulated the actual power. Only time will give the answer. Nothing in history is predictable, and analogies from the past are too inexact to become the basis of confident predictions. Nevertheless, education for a youthful nation, lacking experience, must rely on the study of history so that premature conclusions and hazardous predictions will be discouraged.

We do not know how the power struggle inside the Communist system will develop. Nor can we predict, assuming that overt agreements between the West and Communism are almost impossible, whether tacit agreements may not be reached more and more; and whether such tacit agreements will relieve the tensions, perhaps even limit armaments, if both sides find the burden too heavy.

In short, in facing our hazardous future, we must learn an attitude of both caution and hopeful expectancy, remembering that "if hopes are dupes, fears may be liars." No particular set of academic courses can furnish the new or the old generation with the wisdom sufficient for performing our tasks and solving our problems adequately. But every form of education that brings the experience of the past to bear upon present problems, and that enlarges the realm of insight, will help the new generation to face the unprecedented dilemmas of the day.

HANS J. MORGENTHAU

Sound thinking about political education must start with two propositions: first, that there exists a body of truth which is applicable for all men and all time; second, that professional education is but one among several agencies, and not necessarily the most effective one, through which this truth is communicated. Neither of these propositions is taken seriously by our culture.

The dominant trend in our culture denies the existence, or at least the accessibility to human reason, of objective, universal truth in matters political. Where no objective, universal truth is recognized, no such truth can be taught; and where the political world is conceived as the interplay of ephemeral social forces, the enterprise of education is reduced to learning how to discern these forces and how to avoid getting caught in the interplay. So American political science has been reduced to a technical, descriptive, methodological, and ameliorative enterprise. It has had a predominantly practical orientation during the better part of its history.

Having responded to the ever changing demands of practice in a pragmatic spirit, American political science has no theoretical unity either in content or in method. To pick out at random some courses from two departments of political science with which I am familiar, what has "Plato's Political Philosophy and Its Metaphysical Foundation" in common with "The Politics of Conservation"? Or "Russian Political and Economic Institutions" with "Public Personnel Administration"? The only common denominator that now ties the courses in the political science curriculum loosely together is a general and vague orientation toward the nature and activities of the state and toward activities which in turn have a direct bearing upon the state.

Political science has not generally been able to make the distinction that is a precondition for the development of any true science: the distinction between what is worth knowing intellectually and what is useful for practice. Improvement of the processes of government is still generally considered a legitimate, and sometimes even the only legitimate, element of political science as an academic discipline, to be taught under any of the course headings composing its curriculum.

American political science has been committed largely to the practical goals of liberal reform. But with each supposed achievement in the direction of one of these goals, the old problems have reappeared in a new garb, mocking the scientific pretenses of liberal political science, and yesterday's hope and today's achievement became tomorrow's illusion. To this succession of blows that liberal politics suffered at the hands of experience must be attributed in good measure its disenchantment with reform and its decline in creative thought and action. The grand ameliorative schemes of liberal political science petered out in proposals for piecemeal improvements from which no great things could be expected.

The study of international relations as an academic

discipline has followed a similar course. When, after the First World War, this study gained recognition as an independent discipline, it had three main intellectual interests: history, international law, and political reform. It is not by accident that the first two occupants of the first chair of international politics—founded in 1919 at the University of Wales—were distinguished historians, Professors Zimmern and Webster. Certainly knowledge of history, and more particularly of diplomatic history, forms an indispensable element of international relations, but the intellectual interest of the student of international relations is not identical with that of the historian. Obviously the intellectual interest in international relations is in the present and the future rather than in the past.

The study of international relations has been built upon four subjects: general history, diplomatic history, international law, and a fourth cornerstone, less easy to identify, which is formed by aspirations for a better world, vaguely conceived and identified with whatever remedy seems to be fashionable at a particular time—disarmament one day, arbitration the next, the League of Nations or the United Nations another, collective security still another.

The interstices between those four cornerstones are filled in with an incoherent collection of fragmentary studies ranging the whole gamut of academic disciplines and having only one thing in common: that they transcend the boundaries of a particular nation. At a meeting of the International Institute of Intellectual Co-operation in 1935, one speaker seems to have summarized well the consensus when he said: "One can without doubt call 'international' any phenomenon because it belongs to all countries. Anything one wants to then becomes international. From this point of view, seasickness is an international fact . . . and one can conceive of an international league whose purpose it is to do research and compare the methods with which to combat the disease."

The intellectual unity of an academic field with such a foundation is bound to be of a most superficial nature.

The Philosophic Basis of Political Science

What, then, ought a science of politics, domestic and international, to be like, that does justice both to its scientific pretense and to its subject matter? The answer to this question, insofar as it concerns the scientific pretense of political science, derives from three basic propositions: the importance of political philosophy for political science, the identity of political theory and political science, the ability of political science to communicate objective and general truth about matters political.

Political science, like all science, is a reflection of philosophic propositions. It must be founded upon a philosophic understanding of the nature of man and society and of science itself. It is based upon and permeated by a total world view—religious, poetic as well as philosophic in nature—the validity of which it must take for granted.

During most of the history of Western political thought, the functions of political philosophy and of political science were united in the same persons. The great political philosophers were also the great political scientists, deriving concrete, empirically verifiable propositions from abstract philosophic ones. In our day, political science has divorced itself from political philosophy, and in so doing it has cut itself off from the very roots to which it owes its life, which determine its growth, and which give it meaning. A political science that knows nothing but its own subject matter cannot know even that subject matter well. Contemporary political science, predominantly identified with a positivistic philosophy which is itself a denial of virtually all of the philosophic traditions of the West, has mutilated itself, as it were, by refusing itself access to the sources of insight available in the great philosophic systems of the past.

Why is it that all men lust for power; why is it that even their noblest aspirations are tainted by that lust? Why is it that the political act, in its concern with man's power over man and the concomitant denial of the other man's freedom, carries within itself an element of immorality and puts upon the actor the stigma of guilt? Why is it, finally, that in politics good intentions do not necessarily produce good results and well-conceived plans frequently lead to failure in action; and why is it, conversely, that evil men have sometimes done great good in politics and improvident ones have frequently been successful? Here we are in the presence of the mystery, the sin, and the tragedy of politics. The problems that these questions raise are not scientific but philosophic in nature. Yet without the awareness of their legitimacy and relevance, political science is precluded from even raising certain problems essential to the scientific understanding of politics.

The same anti-philosophic position, prevalent in contemporary political science, is responsible for the present common separation of political theory from political science. Theory, being by definition useless for practical purposes, has been assigned an honorific but ineffectual position in the curriculum, and main emphasis is placed upon science, whose immediate usefulness for society the natural sciences seem to have demonstrated.

Perhaps no event has had a more disastrous effect upon the development of American political science than this dichotomy between political theory and political science.

For it has made political theory sterile by cutting it off from contact with the contemporary issues of politics, and it has tended to deprive political science of intellectual content by severing its ties with the Western tradition of political thought, its concerns, its accumulation of wisdom and knowledge. Aware of this loss, we have resorted to the remedy of adding more courses in political theory to the curriculum, or requiring knowledge of political theory in examinations. However, the remedy has been of no avail.

The very distinction between political theory and political science is untenable. Historically and logically, a scientific theory is a system of empirically verifiable, general truths, sought for their own sake. This definition sets theory apart from practical knowledge, common-sense knowledge, and philosophy. Practical knowledge is interested only in truths that lend themselves to immediate practical application; common-sense knowledge is particular, fragmentary, and unsystematic; philosophic knowledge may be, but is not of necessity, empirically verifiable. What else, then, is scientific knowledge if not theory? It follows that political science cannot be made more theoretical by increasing emphasis upon the separate field of political theory but only by infusing all branches of political science with the spirit of theory.

The same philosophic position that has made political science disparage philosophy and separate itself from theory has also made it deny the existence and intelligibility of objective, general truths in matters political. That denial manifests itself in different ways on different levels of discourse. On the level of the general theory of democracy, it leads to the conclusion that the decision of the majority is the ultimate criterion beyond which neither analysis nor evaluation can go. On the level of analysis of political processes and decisions, it reduces political science to the explanation of the ways by which pressure groups operate and the decisions of government are reached. A political science thus conceived limits itself to the descriptive analysis of a complex of particular historic facts. What a political science of the past has discovered to be true, then, is true only in view of the peculiar and ephemeral historic circumstances of the time, carrying no lesson for us or any other period of history, or else is a mere reflection of the subjective preferences of the observer. The political science of the past is thus reduced, insofar as it seeks empirical analysis, to the description of an ephemeral historic situation.

This being so, contemporary political science is caught in a relativistic dilemma. Either it will be tempted to take flight in a subjective dogmatism that identifies the perspective and preferences of the observer with objective, general truth—thus becoming the ideology of a particular view of society, reflecting particular social interests—or else it will surrender the very concept of objective, general truth, concluding from the subjectivity of its own insights that there is nothing but opinion and that one opinion is as good as another, provided society does not object to it.

If nothing that is true regardless of time and place could be said about matters political, political science itself would be impossible. Yet the whole history of political thought is a living monument to that possibility. Political scientists of the past, reflecting upon matters political under the most diverse circumstances, have developed a common store of truths which are as accessible to us as they were to our predecessors. If it were otherwise, how could we not only understand but also appreciate the political insights of a Jeremiah, a Kautilya, a Plato, a Bodin, or a Hobbes?

A theory of politics, domestic or international, must search for the truth about matters political. In that search it is subject to a purely pragmatic test. Does this theory broaden our knowledge and deepen our understanding of what is worth knowing? If it does, it is good; and if it does not, it is worthless, regardless of its *a priori* assumptions.

Hypothetically one can imagine as many theories of politics as there are legitimate intellectual perspectives from which to approach the political scene. But in a particular culture and a particular period of history, there is likely to be one perspective which for theoretical and practical reasons takes precedence over the others. At one time theoretical interest was focused upon the constitutional arrangements within which political relations took place; in view of the theoretical and practical problems to be solved, this was then a legitimate interest. At another time in the history of political science, theoretical interest was centered upon political institutions and their operations; in view of what was worth knowing and doing at that time, this theoretical interest was again legitimate. Thus political science is like a spotlight which, while trying to illuminate the whole political world, focuses in one period of history upon one aspect of politics and changes its focus in accordance with new theoretical and practical concerns.

The Central Concept of Political Theory

In our period of history, the justice and stability of political life are threatened, and our understanding of the political world is challenged, by the rise of totalitarianism on the domestic and international scene. The novel political phenomenon of totalitarianism puts in doubt certain assumptions about the nature of man and of society which we took for granted. It raises issues about institutions which we thought had been settled once and for all. It disrupts and overwhelms legal processes on which we had come to look as self-sufficient instruments of control. In short, what has emerged from under the surface of legal and institutional arrangements as the distinctive, unifying element of politics is the struggle for power—elemental, undisguised, and all-pervading. As recently as the end of the Second World War, it was still held by conservatives, liberals, and Marxists alike either that the struggle for power was at worst a raucous pastime, safely regulated by law and channeled by institutions, or that it had been replaced in its dominant influence by economic competition, or that the ultimate triumph of liberal democracy or the classless society, which were expected to be close at hand, would make an end to it altogether. These assumptions and expectations have been refuted by the experience of our age. It is to the challenge of this refutation that political science must respond, as political practice must meet the challenge of that experience.

Yet while political science must come to terms with the problem of power, it must adapt its emphasis to the ever changing circumstances of the times. When the times tend to depreciate the element of power, it must stress its importance. When the times incline toward a monistic conception of power in the general scheme of things, it must show its limitations. When the times conceive of power primarily in military terms, it must call attention to the variety of factors that go into the power equation and, more particularly, to the subtle psychological relations of which the web of power is fashioned. When the reality of power is lost sight of, it must point to that reality. When law and morality are judged as nothing, it must assign them their rightful place.

It may be pointed out in passing that all great contributions to political science, from Plato, Aristotle, and Augustine to *The Federalist* and Calhoun, have been responses to such challenges arising from political reality.

They have not been self-sufficient theoretical developments pursuing theoretical concerns for their own sake. Rather they were confronted with a set of political experiences and problems that defied understanding with the theoretical tools at hand. Thus they were compelled to separate in the intellectual tradition at their disposal that which was historically conditioned from that which was true regardless of time and place, to pose again the perennial problems of politics, and to reformulate the perennial truths of politics, in the light of the contemporary experience. This has been the task of political science throughout its history, and this is the task of political science today. There is, then, in political science what might be called a "higher practicality," which responds to practical needs not by devising practical remedies but by broadening and deepening the understanding of the problems from which the practical needs arise.

A central concept such as power provides a kind of rational outline of politics, a map of the political scene. Such a map does not provide a complete description of the political landscape as it is in a particular period of history. Rather it provides the timeless features of its geography distinct from their ever changing historic setting. Such a map, then, will tell us the rational possibilities for travel from one spot on the map to another, and which road is most likely to be taken by certain travelers under certain conditions. Thus it imparts a measure of rational order to the observing mind and, by doing so, establishes one of the conditions for successful action.

A theory of politics, by the very fact of painting a rational picture of the political scene, points to the contrast between what the political scene actually is and what it tends to be but can never completely become. The difference between the empirical reality of politics and a theory of politics is like the difference between a photograph and a painted portrait. The photograph shows everything that can be seen by the naked eye. The painted portrait does not show everything that can be seen by the naked eye, but it shows one thing that the naked eye cannot see: the human essence of the person portrayed. Thus a theory of politics must seek to depict the rational essence of its subject matter.

A Theoretical Foundation

The application of these general reflections to international relations leads to the postulation of certain principles, only the most basic of which can be mentioned here.

Underlying all political education must be the awareness that all the specific manifestations of a particular culture contain an element of universality, however undiscoverable or unprovable it may be in a particular instance. Political education, then, must take into account an element that transcends the limits of any particular culture. The element of universality, transcending any particular culture and common to all, may be called human nature. However different its specific manifestations at different times and places, it is the same everywhere and at all times. If we did not assume this identity, we could see in other cultures, past or present, only a mass of incomprehensible facts or a distorted image of our own culture. Thus every historian and political scientist must assume implicitly the identity of human nature in time and space in order to be able to understand at all, however loudly he may deny its existence. At this point we come face to face with what is perhaps the most serious shortcoming of the contemporary study of international relations.

This shortcoming is the assumption that the key to understanding a culture lies in investigating its specific attributes. If we want to understand China, we must study

China; if we want to understand France, we must study France. Yet may it not be said that, in order to understand China or France or any other complex of specific phenomena, it is first necessary to understand mankind, of which all specific phenomena are but particular manifestations? If I know something about human nature as such, I know something about Chinese and Frenchmen, for I know something about all men. It is true that this something I know about all men is likely to lead me astray if I try to explain through it the concreteness of a particular historical situation. Yet without such a conception of human nature, made articulate in a philosophy of man and society, international relations cannot be understood in a theoretical manner.

That this is not mere idle speculation everyday experience shows. Why is it that I am able to understand the Homeric heroes or Chaucer's pilgrims without having mastered the area research of ancient Troy and medieval England? Why is it that I am able to comprehend the domestic and foreign policies of contemporary Russia without being an area specialist in the Russian field? Why is it that I have a general understanding of contemporary China while I am virtually ignorant about China as an area? Why is it that the members of the British foreign service have been traditionally trained in the humanities and more particularly in the classics and then sent in succession to the four corners of the earth, showing frequently superb understanding of the areas in which they worked? The answer to these questions has already been given: If you know something about man as such, you know something about all men.

International relations as an academic discipline requires a hierarchy of intellectual interests, one of which is predominant, providing the principle of integration, while the others are subsidiary, supplying the knowledge necessary for the satisfaction of the predominant interest. What is the predominant interest of international relations as an academic discipline? Two different answers must be given to that question. One is that the possible predominant interests are as numerous as are the legitimate objects of intellectual curiosity. It is, then, as legitimate to put economics in the center of international relations as it is to put law or geography there, and to subordinate other disciplines to the predominant economic, legal, or geographical interests. In this view, many "sciences" of international relations are possible.

The other answer to the question of what the predominant interest of international relations must be assumes that among a number of legitimate interests there is one that demands special attention in a particular period of history. The educator must ask himself which among the many possible foci of international relations is most important for the student's interest to center upon, and the scholar must ask himself which among the many perspectives from which one can investigate international problems is most important from the theoretical and practical point of view. Today most institutions and students have turned to the study of international relations because of their interest in world politics. The primacy of politics over all other interests, in fact as well as in thought, needs only to be mentioned to be recognized. The recognition of this primacy of politics cannot but lead to the suggestion that, among the legitimate predominant interests upon which international relations as an academic discipline might be focused, international politics should in our time take precedence over all others.

Political Leaders as Educators

While it is the task of political education to communicate the truth about matters political, it is an illusion, to which professional educators are prone, to believe that the success of the educational enterprise depends primarily upon the quality and quantity of professional education. This is not necessarily so in any field of education, and it is not even typically so in the field of political education.

It is an obvious fact of experience that professional education is only one—and not necessarily the most important—among several factors that mold the mind and character of those to be educated. Family, society, the experiences of life itself are more persistent and authoritative teachers than the schools. Education, to be effective, must be organically attuned to the totality of educational influences to which the individual is subject.

This organic relationship between professional education and the totality of educational experience is strikingly revealed in the field of political education. The measure of success that education in world affairs has had in the United States in recent years is primarily due not to the quality and quantity of its professional manifestations, but to the experiences that the American people have undergone during and after the Second World War and to their interpretation by political leaders. What is being taught today in American colleges, say, about the balance of power, to be accepted almost as a matter of course, was taught, however sporadically, thirty and twenty years ago, only to be dismissed as absurd. It is not professional education that has made political understanding in this respect possible. Rather it is political experience that has made the truth plausible. Professional education has proved the validity of political experience through historic example and analytic demonstration. Without that experience, however, political education would have remained as ineffective as it was before, while without professional education political experience would have remained unconvincing and inarticulate. What has been said of the three stages through which all truth must pass applies with particular force to political truth. First, people dismiss it as impossible; next, people dismiss it as immoral; finally, people accept it as self-evident.

This achievement of political education is the task of political leadership. Only those whom the people have elected because they have confidence in them, or those in whose judgment the people otherwise confide, can make the truth of political experience explicit. For only they have the authority to gain acceptance for a political truth which is not self-evident from the outset. What they need is the political judgment to see the truth and the political courage to tell it. Thus the most effective political educators in America have been the great Presidents, Senators, and commentators.

For their authority there is no substitute, either in professional education or elsewhere. Nor is there elsewhere a real substitute for the other two requirements: political judgment and political courage. Professional education may supply them in rather rare instances. But without the authority of the political leader, the voice of professional education does not carry far.

⁂ *Discussion*

NAGEL: I would like to understand Professor Morgenthau's position a little more clearly. May I ask him to spell out what some of the fundamental principles of political science are?

MORGENTHAU: One of them is the concept of power, which I take to be as central a concept in political science as that of "value" in economics.

Take, for instance, the principle of the balance of

power. I have always regarded the balance of power as a principle as fundamental for politics as the law of gravitation for physics. The principle says that, where there are a number of autonomous units anxious to preserve their autonomy, the only policy they can pursue to that end is a policy of the balance of power, that is to say, of pitting their own power against the power of the other side. On the international scene this leads to an armament race, by which one side tries to keep up with and overtake the other side, and vice versa.

There have been periods in the history of our thinking on foreign policy, for example in the 1920s and 1930s, when this contest was regarded as something to be abolished, and when it was assumed that statesmen had a choice between a foreign policy based upon it and a policy based on some other principle. Even during the last year of the Second World War, and immediately afterwards, there were many who believed that the United Nations provided an alternative to power politics and to the principle of the balance of power.

Implications for Education

HOOK: I would like to ask some questions which bear upon education and our failure in education. I would take as my point of departure a thing that startled me some years ago. In about 1945, *Fortune* had a poll in which it asked the following question: What do you think the prospects are of peaceful cooperation and coexistence—though they didn't use the word "coexistence"—between the United States and communist Russia? The overwhelming majority of people in the universities maintained that the probability was very good. The overwhelming majority of workers, of people drawn from the trade-unions, and farmers, maintained that the probability was very small.

Now our knowledge of guiding laws is tested in part—not altogether—by the ability to make predictions. I have sometimes wondered what the basis was of this terrible error in prediction.

I remember attending a conference at the Princeton Bicentennial in 1947 in which many said, "Understanding is a necessary and sufficient condition for the establishment of peace." Somebody in the audience raised the question whether that was always true, whether, if you understood the way of a snake, you would necessarily get along with it better, whether there weren't some things which you opposed the more strongly, the more you understood them. Wasn't it true that if we had understood Hitler better, we would not have appeased him? Well, the individual who raised the question was regarded as a red-baiter and a disrupter.

Harold Laski had just come away from Russia with the news that Stalin had told him personally that all the Soviet Union wanted was peace. His statement was applauded to the echo, and any one who disagreed with it was regarded as ignorant in this field. Now my question is: What happened to American education which made so disastrous a misunderstanding about Russia possible among educated men?

My own hypothesis is that the error arose from an attempt to understand political reality in terms of principles of power only, overlooking the unique historical situations created by the rise of totalitarianism, both of the Nazi and the Soviet variety. Let me put it another way. I came to the tentative conclusion that the great errors, both in American and English foreign policy, were the result of a failure to understand politics based on a *Weltanschauung*. The Russians maintain that so long as capitalism exists it inevitably breeds war; believing that they can't possibly believe in coexistence. Differences among the Western nations are power differences and economic differences, but between us and Soviet Russia there is a more profound difference in ideological *Weltanschauung.* The Communist outlook is so foreign to our conception of politics, we of the West have so much more in common with each other than separates us, that, in a sense, we can no longer interpret the political scene. I put it to you then, Professor Morgenthau, don't you think that our education in this respect has been deficient, in the sense that we have not taken the ideas of others seriously enough?

MORGENTHAU: I agree that our education has been disastrously deficient, but not quite for the reason you indicate. The reason is rather to be found in two types of misunderstanding which dominated universities.

During the interwar period of the 1930s, students were unshakably convinced that munitions makers and Wall Street were responsible for war; it was a period when the findings of the Nye Committee were the expression of the prevailing philosophy. In other words, foreign policy was not recognized as an autonomous human activity; it was regarded by this school of thought as a mere manifestation of economic causes. Another expression of this view was the "have-not" theory. All trouble in international affairs was supposed to stem from disparities in wealth: if you had only given a loan of ten billion dollars to the Nazis and the Japanese, they would have been satisfied, and would not have gone to war.

I don't need to tell this group that the exact opposite is true: that under modern conditions only prosperous nations can think of war, for poor nations cannot wage war.

The other misunderstanding, maintained by a still more influential school of thought whose most eminent and eloquent spokesman was Woodrow Wilson, was the conception that international troubles arose out of fortuitous circumstances—such as weaknesses of governmental systems (e.g., lack of democracy) and suppression of the national aspirations of peoples. Let those problems only be solved, and the problems of foreign policy would go away and we could forget them.

Part and parcel of this school of thought was the theory to which you referred, Professor Hook: that conflicts are merely the result of misunderstandings. As you have indicated, in international relations the exact opposite is frequently true. It is because we understand the Russians only too well that we have trouble with them. As long as we didn't understand them and deceived ourselves about them, we didn't have much trouble; we rested on the assumption that we could charm Stalin into cooperation.

To come to the answer that you yourself have given, that this misunderstanding of the fundamental facts of political life is the result of our not taking ideas seriously enough, I think that is only peripheral to the actual problem. It is rather that we thought that Hitler and Stalin were essentially people like ourselves, as civilized, as modern, and, in particular, as rational. Hence the genuine shock to which Neville Chamberlain confessed on his deathbed, when he said of Hitler: "I trusted this man, and he deceived me." The same shock occurred on a collective basis in the United States when we realized that Stalin had deceived us. I don't think it is so much a question of these men's ideas as of their character.

International Thinking in the Schools

WOODRING: Mr. Morgenthau has expressed his dissatisfaction with the way our people are educated on these problems of international politics. I think we all share his dissatisfaction. But I would like to ask him what changes in the contents of either secondary or higher education would best serve to correct this defect?

MORGENTHAU: I have seen some projects in high school which impressed me considerably, not because of their technical excellence, of which I can hardly judge, but because of their awareness of the basic philosophic problems to which I have referred. There has been a real revolution of late in our thinking in foreign policy. I am, perhaps, a living example of this change. I haven't changed my own opinions, but certainly when I expressed these opinions before the war, I was completely ineffectual. When I made a speech in Milwaukee in 1943, in which I said that after the war there would be another balance of power, and in all probability the Soviet Union would be our antagonist rather than Nazi Germany, somebody said, "You are a fascist agent." I doubt whether anyone would say that now. There is an acceptance today of certain basic insights into the nature of foreign policy which would have been unthinkable even fourteen or thirteen or twelve years ago. So there has been a real revolution, under the impact of the experience of world events and the interpretation of those events by a small elite of leaders of political opinion in the United States.

The only advice I can give is of a general nature, and is in substance what I said before, that instead of running after current events, and the headlines, and the interpretation of yesterday's *New York Times* in class, there ought to be an attempt at systematic understanding, not of what De Gaulle is all about, let us say, but what foreign policy and politics in France or anywhere else is all about. In the light of this over-all understanding, the students will be able to grasp and put into a framework the events of today and tomorrow.

BOEHM: During President Eisenhower's campaign for re-election in 1956 unfavorable events broke in the Near East just three or four days before the election, but according to press polls, to the voters they made no significant difference.

MORGENTHAU: That is correct. The voter has no interest in foreign policy *per se,* except as an event seems to impinge upon the question of war and peace for his own country. As long as only the Lebanese shoot each other, this is no problem for the voter in Kansas.

TUMIN: How do you know that?

MORGENTHAU: Aside from such experience as I have had, I know it from the intensive interviews which Harrison has made, for instance, concerning the attitude of people toward foreign policy. The people had not the slightest interest in anything except employment, prices, and the general problem of peace or war.

TUMIN: Mr. Morgenthau, in national polls taken four times a year in recent years a random sample of the population has almost invariably named foreign relations and foreign policy as number one on the list of most important problems.

MORGENTHAU: Which raises the question of the significance of such polls for the actual convictions, and more particularly the intensity of the convictions, held by those people. If somebody comes to my office with a questionnaire, and asks me a set of questions, I might give one set of answers, and my actual convictions, and more particularly the intensity of my convictions, might require a quite different set. An experience of my own shows this clearly. Some time ago a very intelligent pollster came to my office and asked me about my attitude toward the quarterly reports of corporations in which I hold a very small number of shares of stock. It was a most interesting question, and I gave serious and, I thought, competent answers to it. After the lady had left, it occurred to me that the only thing I do when I get such a report is to look at the earnings, compare them with the quarter of last year, and then throw the thing into the waste basket. My answers really meant little, because I am neither a serious nor a competent investor.

So I think one ought to be very careful before one draws conclusions as to the intensity of people's interest in international affairs, political or intellectual.

People are interested in problems which affect them directly, and only to a sharply decreasing extent in problems which affect them only remotely. Their concern is by and large proportionate to the degree in which their interests are affected. Foreign policy, such as the problem of the Middle East, is far removed from either the interests or the knowledge of the individual. How many voters can point out Lebanon on a map? So I think there is enough evidence to justify saying that the interest of the average voter in foreign policy is very small.

TUMIN: How can we generate more interest?

MORGENTHAU: I don't believe you can generate it short of showing to the individual—and this is a tenuous process—that his interests are somehow tied up with the problems of the Middle East, Africa, and the rest.

KUSCH: I should like to know explicitly what you would recommend on various levels of instruction from kindergarten through college.

BOEHM: One of our summer schools, looking into the content taught in elementary schools, found that except in geography, the textbooks and library materials rarely relate to anything outside our own country—even to South America or our two neighbors, Canada and Mexico.

HOOK: As far as elementary school is concerned, we need to give the students the sense of the unity of civilization in a diversity of cultures. We need to substitute books which would in some way negative the natural chauvinism and nationalism of the young. Their tendency is to thrill at the sight of the pictorial statistics which show the United States as the greatest of all countries in producing coal. I remember as a child being impressed with that, and taking a vicarious satisfaction. It was innocent enough; but if it isn't countered by their realizing that even in enemy countries the people are fundamentally human beings and not really our enemies, we lose something important.

On the high-school level, can't we pay more attention to the way different people think? On the college level, shouldn't there be a requirement for the systematic study of totalitarianism, Communism and Fascism, not in a doctrinaire or dogmatic spirit, but with the aid of the text the Russian students use, so that our students can see what the Russian students believe, and believe about us. Can't our students be given a sense of the geography of the mind? Of course, that can't be done in elementary schools. But for others the educators must have thought of many approaches.

BOEHM: In Pennsylvania we tried in the current year to put in the secondary schools a required year's course on world cultures and their histories. As soon as the vocational teachers got wind of it, we had to go back on our tracks. History has been integrated with geography in the eighth, seventh, and even fifth grades in many places; but, by and large, what our students know about the culture of the Chinese, for example, is very little.

HOOK: Why did the commercial teachers object to the proposed course?

BOEHM: They said that they can't carry through their full program as it is, and that if you introduce any more mandatory subjects, they can't complete the schedule. Now the demand for science has come along, and we are threatened with losing even geography.

MORGENTHAU: In high schools, I think one of the indispensable requisites is the study of history. In the last ten years a great deal has been done to remedy the terrible student deficiency not only in the understanding of history but in the feeling for historic sequence. That the Crusades came before Luther, and that Luther was before Napoleon, is something which you and I regard as self-evident. But students frequently, in my experience, had to learn this by heart, and weren't quite sure who came first. I remember that when I taught in Kansas City in the late 1930s in my foreign relations course I always asked the question: How did the United States acquire the Philippines? And any answer was possible, from Washington down to Roosevelt.

Teaching history certainly is necessary, particularly American political history. I think there is great danger in those world history courses which give a stratospheric view of the history of the world in thirty hours or so. I would rather recommend an intensive study of the political history of the United States, and more particularly the history of its foreign policy.

VAN DEN HAAG: I cannot imagine that, with all possible gimmicks, and the most ideal educational methods at our disposal, we will ever find a good enough way of teaching in high school to lead the high-school student to have really correct opinions on foreign policy. Frankly, I think that is asking the impossible. I would be delighted if he had correct opinions on grammar and if he knew some history, which is no doubt necessary for correct decisions on foreign policy, but surely not sufficient. As for really sufficient knowledge, I would be happy if he got half that much in college. Aren't we losing what we could do in favor of something impossible, when we set ourselves such a task?

MORGENTHAU: You pose your question in a way which allows only a negative answer when you ask whether we can equip each student to make *correct* decisions. How do you teach the President and the Secretary of State to make correct decisions? How do you teach anybody to make such decisions? We are always beset by doubts—even great statesmen have been. This is the reason some statesmen have taken refuge in astrology—seriously, they have; they needed some kind of certainty.

Politics and Law

BUCHANAN: In our common schools, to use Mr. Child's phrase, there is a tradition of teaching civics. I remember this, again, in my little school in Vermont. I didn't know quite what it was about; it was a boring course; and we did it in a catechistic way. I didn't learn very much. I knew how town meetings ran, because I used to attend them. I came to have a kind of patriotic sense of American political feeling. I believe this does not exist any more, or in very small measure.

Well, in the last ten or fifteen years I have become very much interested in law. And—this is the point—I can imagine a very good course in law which would be a combination of civics and jurisprudence for students in high school, though of course it wouldn't be professional law at all.

The ignorance of the American public about the simple legal arguments they are up against every day is appalling. There is a fear of government in everybody's mind at present. The government is away off there, and anything it does is an invasion of liberties. This is an extraordinarily shortsighted view, and I wonder if something can't be done about it. It seems to me it would be entirely possible to teach the sort of course I have in mind. I know two or three young lawyers who would be much interested in doing it.

This leads me to a bigger question, one that I thought Mr. Morgenthau might be alluding to in his paper. Some years ago I ran across a quotation from Einstein, though I can't recall the context. He said: "The only way to think about human destiny in our time is in terms of politics." This suggests that some of the humanistic bases used in the Renaissance are *passé*; they are no longer available to us. I am wondering if there isn't a new kind of epoch opening now, in which politics may be the great unifying subject —law and politics. Mr. Morgenthau seems to me right in saying that we have never studied political science proper in this country since the founding fathers; we have forgotten everything in the Federalist papers and the other great documents. But these have a reflective significance for all the problems facing us. Do you see anything in what I am saying?

MORGENTHAU: I would have some doubts about your proposal of a law course. But I have a direct professional interest in making politics a central topic of intellectual endeavor.

NAGEL: Dr. Morgenthau, the suggestion you made about the teaching of history I find a little opaque. Of course we have been exposed to the teaching of history for a long time. I took history in high school in the middle of the first World War; and what we studied we justified as an attempt to understand the conditions, causes, and objectives of the struggle. So I would like to go back to the question: What is the content of political theory and political science? I still don't understand what these principles are in terms of which history is going to be presented. You suggested the principle of the balance of power. But would you say that this is a key which will unlock all puzzles about historical development? Surely there are others.

MORGENTHAU: I can imagine a high-school course which would present the history of American foreign policy in terms of the defense and promotion of the national interest of the United States, conceived in terms of what is fashionably called power politics. One could show the conception that the founding fathers had of the nature of American interests, and of the relations between the United States and the rest of the world; one could show the policies they advocated in order to defend and promote the interests thus conceived.

Then one could show how this basic conception underwent drastic change—how it changed in 1898, and how Woodrow Wilson put forth a philosophy fundamentally hostile to this original conception of the founding fathers. Finally you could show how the experience of the Second World War led to a rediscovery of the eternal verities that had been expressed by the founding fathers, and perhaps suggest, however imperfectly, how those objective truths might be applied to contemporary problems.

Educating for Citizenship

DIEKHOFF: I would like to return to Mr. Vandenhaag's question as to how the high school or any other school can prepare citizens for wiser decisions. There are two things that we may perhaps fairly ask the schools to do. Mr. Morgenthau has told us that the politicians are more effective teachers than professors or high-school teachers. We can ask the school to prepare people better to understand and learn from the politicians. We can ask them also to help people learn from events. But in all our discussions we seem to have been assuming that school is terminal. It seems to me we are too much inclined to ask the school to finish the work, whereas in the business of making decisions, the citizen must learn on the job, as in other jobs.

BUTTS: One way of putting it is whether we should emphasize in our schools a systematic body of theory and organized knowledge, or whether we should stress relating theory to practice, to the making of judgments and decisions. It is clear that Mr. Morgenthau prefers the theoretical and systematic approach to one that has to do with current and practical issues. The trouble with political science, if I understand him rightly, is that it has been dealing too largely with the immediate issues before the American people.

I hold to the alternative that it is our business to improve the quality of judgments. I think we should start as early as we can to give students practice in making judgments; and we should try constantly to refine the kind of judgments that they make. Their skill in making them can be formed through dealing with the headlines and current events, though their judgment will not be properly refined unless it is based on a thoroughgoing study of foreign policy and the balance of power. I don't see why we should confine the students in high school to the study of the Federalist papers. I think they should study them, but shouldn't they also give vitality and meaning to the principles by considering them in relation to Little Rock, which raises as many fundamental questions about balance of power, federal, state, and local, about courts, the military, and the whole range of issues regarding citizens and their government as the founding of the Constitution itself?

BESTOR: I don't think there is any contradiction here; it is a question of emphasis. I would also like to suggest this: the contemporary problem isn't necessarily the only useful one. It is perfectly possible to treat the problem of the formation of the Constitution in a way which would illustrate and stimulate practical thinking. In fact it has some substantial advantages, since the returns are now all in.

BUTTS: Are they? Little Rock raises these problems over again.

BESTOR: Even granting that, I am not sure that, pedagogically speaking, to take an unsolved problem and deal with it without being able to validate one's conclusions is the only or the best procedure.

But to turn to a somewhat different point, it seems to me that our discussion illustrates one of the things that have made our quest for educational improvement almost frantic. The questions directed to Dr. Morgenthau suggest that someone should find a panacea and quickly propose it. I suspect that Dr. Morgenthau thinks, as I do, that all this underlines the importance of doing better the things we are doing already. The study of history has not been revolutionized by any new discoveries, and in my judgment is not likely to be. It is to be improved not by altering its general character or finding new gimmicks for it, but by eliminating some of its present failings.

KENNEDY: As some of you know, at Amherst we have a required course for sophomores called Problems of American Civilization. This course starts on a chronological basis, and then in the second semester reaches contemporary issues. We have considered Little Rock from the constitutional point of view. Professor Morgenthau himself has lectured in this course on Wilson and the Treaty of Versailles. We have a problem entitled "The Declaration of Independence and the Constitution." There is another problem on imperialism, which focuses on the Spanish-American War. So we have a variety of problems of foreign policy as well as of domestic policy in this course.

I think a similar course could be used at the high-school level, though you would have to use different readings and scale down the intensiveness. But I see no reason why you couldn't use case methods, or combine chronological and contemporary problems, in such a way as to give the student a sense of historical development and also bring him to contemporary issues at the end.

BOEHM: We have had that in the Pennsylvania schools for thirty years.

ALLEN: In the State of New York we have the so-called "problems of democracy" course which is a watered-down form of this.

HOOK: How is it working out? I hear that sometimes it is good, sometimes not.

ALLEN: That is about the truth, I guess.

TUMIN: Isn't the one clear conclusion from our discussion that the indispensable course is a course in how to think? Fifteen years ago most high-school students in certain sections of the Northeast were taught an interpretation of the founding of the Republic, provided by Charles Beard, which is now being systematically denied by the historians of the Constitution. If this is so, if there can be major reversals in historical interpretation on the basis of new evidence, and if our students forget much of the factual content of our courses anyhow, wouldn't we be well advised to teach them, at the earliest possible level, how to read critically and how to think critically? Would not this stand them in better stead than absorbing a knowledge of events about whose meaning and connections we have to remain doubtful?

New Courses or Better Teachers?

WOODRING: It seems to me that there is a tendency here to avoid facing the real problem. If you can't teach secondary-school students to make political decisions, I think you are saying democracy can't work. Most of these students won't go to college, yet they have to vote.

One unrealistic proposal is that you solve the problem by teaching everybody American history. But our high schools have been doing that since the beginning of time, though apparently not well enough. This doesn't seem a very good solution.

Another one is teaching a problems course. This has

been experimented with for twenty or thirty years, with results which in some places are disastrous and others magnificent, depending on the teachers. So where we seem to stand is on the fact that we must get better teachers for these areas. Tell us how to do that, and you have the problem licked.

HOOK: What needs stress may not be so much what courses should be taught as a fuller awareness of the importance of the American tradition of freedom, which stands in my mind as the abiding expression of the American experience. For all our isolationism in the 1930s, students had an interest in freedom, awakened perhaps by the danger of Fascism, which I find lacking among them now. It needs to be kept awake. Perhaps it is emphasis of this sort in our courses that is to be desired, rather than new courses themselves.

MORGENTHAU: An enormous contribution could be made to political understanding if American history were taught not simply as a recital of facts but as a manifestation of certain fundamental principles of politics, and as a manifestation of a universal human experience. For such an approach would show that the problems which confront us today in foreign policy are not the result exclusively, or even primarily, of a particular configuration of weakness and wickedness in some individual or group of individuals, but that they are rooted in human nature, that they are the result of unvarying configurations in human interactions. I think a student who saw this would have a much deeper insight into the political problem he confronted in the future. I am reminded of Jacob Burckhardt's statement that history should teach us not how to be smart for one day but how to be smart forever.

NAGEL: I would certainly endorse everything that Dr. Morgenthau has said about the importance of teaching certain fundamental political principles, if these principles could be discovered. But I am not convinced that there are any such principles, comparable to principles that have been established, for example in physics, chemistry, and biology. In the history of the human enterprise, and in the social sciences, there is no unified body of theory in terms of which you can hope to understand the complexities of any one situation.

I would agree, Dr. Morgenthau, that there are cases where viewing the situation as a struggle to maintain the balance of power is illuminating. In other cases, the Marxist interpretation would seem more illuminating. In still others, the sort of thing that Dr. Hook has emphasized, namely, a passionate desire to preserve freedom, seems to give us a better key. But to place the whole of history on the bed of Procrustes of some one allegedly definitive political truth or some one set of these, seems to me to be doing a serious disservice to political science.

MORGENTHAU: I can only reiterate what I have said before: if there were no general truths about foreign policy, at least of an implicit kind, you would be unable to understand or learn from such a writer as Tacitus. Without the assumption of a set of political truths, which are true regardless of time and space, you couldn't hope to understand history.

When Democrats Meet Communists

JACOBS: I teach a course on international affairs in the adult division of New York University. This year I had two Soviet citizens in my class, members of the Soviet diplomatic service. Apparently they registered in order to get a sense of what goes on in this kind of class. I would say they are of more than average intelligence, though not particularly bright. What has struck me about them is how well they are educated—or perhaps I should say indoctrinated—the resourcefulness with which they express their point of view and meet the problems raised.

I ask myself, how it is that we don't turn out students equipped to cope with their position as well as they handle ours. Why was it so easy for the Chinese in Korea to brainwash American soldiers? Our G.I.'s had no kind of basic *Weltanschauung*. Perhaps we don't want that. But there must be a way, short of indoctrination, of enabling our students to meet the challenge of the products of the Soviet educational system.

BESTOR: Sometimes we overlook the importance of basic common knowledge.

One of the things that underlies a common *Weltanschauung* is a common range of knowledge. And before one can draw generalizations—analyze constitutions or conflicts of power—there must be an underlying fund of historical data. I think we ought to be ready to say that certain things should be taught as the necessary foundation for dealing with political problems. In history, and I should think in literature as well, a great deal is to be gained simply by acquaintance with this material, on which all kinds of theoretical constructions can be based and by which they can be tested. I don't think Dr. Morgenthau suggested that the theoretical analysis precedes the data; it assumes the data as an underlying foundation. I think there is a great deal of wisdom in the view, which seems to prevail in other educational systems, that one begins a study of history at a fairly simple level, that cumulatively it is raised, expanded in time, and extended in depth, and that it provides the foundation for later discussion, on the theoretical level, of political science, economics, and sociology. I think we need more of the stability that concentrates on improving such admittedly central subjects as history and spends less energy searching for panaceas.

MORGENTHAU: I would like to address myself to the two questions that Mr. Jacobs raised. First, concerning the two Russian students. There is no denying the dialectic advantage which a Soviet citizen has under the circumstances. Schopenhauer gave an answer to the question when he asked why it was that Marxist materialism, though completely at odds with the known facts, has had such an impact on the Western world. His answer was that here is a simple doctrine which seems to explain an enormously complicated set of facts with a formula intelligible to a child in the street. Against this kind of dogmatic thinking, which is able to shift its position continuously and is always right within its own terms, there is no easy defense. Reason, based upon experience, is at an enormous disadvantage when it is confronted with this simple-minded and intellectually dishonest method of arguing.

Mr. Jacobs' other problem, concerning the weakened loyalty on the part of American soldiers, arises as a result of the very lack of political awareness and education that we have been discussing. The trouble can be remedied, in part, by formal education, though I think the really decisive contribution must be found in political leadership.

VAN DEN HAAG: You probably noticed that the Turkish soldiers who were taken prisoner at the same time seem to have resisted easily the things that American soldiers found it difficult to resist. I am quite unwilling to attribute that to their superior education. Hence, I believe that education is almost irrelevant to that particular problem.

MORGENTHAU: You are correct that the superior attitude of the Turkish soldiers was not due to education, but to the particular social integration within which they lived and worked, and which we would not want to imitate. Nevertheless, I would not go so far as to say that education isn't relevant. A G.I. who is informed about the mission he has to perform, and what his country stands for in this situation, is less likely to fall prey to brainwashing than a soldier who is unaware of what it is all about.

✻ Douglas Bush

✻ Ernest Nagel

✻ and Panel

4

Science and the Humanities

DOUGLAS BUSH

While other subjects in the curriculum may have their ups and downs, it is always proper to speak of "the plight of the humanities," and in the hushed, melancholy tone of one present at a perpetual deathbed. For something like twenty-five hundred years the humanities have been in more or less of a plight. I should like to give some reasons for the uphill struggle they have always had and always will have. The chief reasons may be grouped under the heading of original sin. Since the humanities are opposed to and opposed by man's animal nature and animal drives, the essential causes of opposition are very old; but they have taken on many new forms and have gained enormously in strength in modern times.

First of all, there is the inspiring and dispiriting fact that, while the goals of science are predictable and attainable, the goal of the humanities is not—unless human nature undergoes a miraculous transformation. It is much easier to make discoveries in the laboratory, to put satellites into the earth's orbit, than it is to achieve humane wisdom, imagination, and insight in thought, feeling, and action. And the latter, the true end of man, is far less compelling than lower ends. The mass of mankind has always been mainly absorbed in the struggle for subsistence or comfort or pleasure or power. To recall a saying of the wise humanist Montaigne, Socrates was a greater man than Alexander because, while Alexander conquered cities, Socrates conquered himself. In the Renaissance creed, man was a creature halfway between the beasts and the angels, drawn both downward and upward by elements in his own nature; the aim of education was to make him less like a beast and more like an angel. The aim of the humanities, an aim that can never be fulfilled but can never be abandoned, is to humanize and civilize the aggressive and sensual animal, to lead him to realize his distinctively human endowments, to refine and multiply his moments of vision, to free his better self from bondage to his ordinary self.

Sophisticated Vulgarity

No one would disparage the multiplying benefits that science and technology have supplied, but it cannot be denied that they have also imposed stultifying pressures on man's mind and character and aesthetic sensibility.

Whatever may be hoped for from science in the distant future, it would seem that at present we have more to fear from the mass civilization of our Western world than from Russia. The modern rise in the material well-being of the common man—which includes woman and child—of course has been long overdue, but one unhappy result so far has been the debasement of traditional culture.

> All things are a flowing,
> Sage Heracleitus says;
> But a tawdry cheapness
> Shall outlast our days.

The general diffusion of literacy has enabled millions to consume the husks provided for them and to make more precarious than ever the survival of high standards of enlightenment and taste. As Reinhold Niebuhr remarked a short time ago: "A technical or technocratic culture generates powerful utilitarian pressures and develops means of mass communication—movies, picture magazines, tabloids, radio and television—which all tend to reduce the culture to a kind of sophisticated vulgarity beside which the vulgarity of the unlettered man of other ages will appear as pure innocence."

This sophisticated vulgarity is not a completely new thing; what is new is that it now makes an immediate impact upon hundreds of millions of people. Nor is it a purely American phenomenon, though foreigners often complain that it is. But in this as in other fields of endeavor, Americans have greater resources. Those who operate mass media may be assumed to have a shrewd and solidly verified knowledge of what the bulk of the public wants, and they evidently set the mental age of that public at about fourteen. Good things, to be sure, do appear now and then in most of these mass media, but they so commonly fail to win attention or approval that the commercially minded producer is not encouraged to try again. For obvious evidence concerning popular taste one may look at any newsstand or listen to radio and television programs or contemplate the stream of big and gaudy automobiles—which, even if not paid for, presumably inflate the egos of their owners. If the tone, the crass and blatant tone, of mass civilization has not been largely created by advertising, at any rate advertising is its sufficient symbol.

Moreover, the media and standards of mass civilization do not operate merely on the lowest economic and cultural levels; they more or less infect the millions who have attended college and might be thought to have higher interests and better taste. One symptom is the way in which the uneducated misuse of words and idioms becomes, overnight, common usage among the supposedly educated and promptly gets into the dictionaries. It is a painful fact of American life and education that, after leaving college, the majority of people virtually give up serious reading; and that, of the minority who do not, only a fraction pays any attention to literature, new or old. In a survey made in 1956, only 17 per cent of American adults were found to be reading books, as compared with 31 per cent in Canada, 34 per cent in Australia, and 55 per cent in England; 57 per cent of our high-school graduates and 26 per cent of our college graduates had not read a single book during the preceding year; of the college graduates, 9 per cent could not name the author of any one of twelve famous books in English, and 39 per cent could not name more than three; 45 per cent could not name any recently published book.

According to Alfred A. Knopf, book publishing has come to be dominated by the gamble on the smash hit, which depends on mass appeal. We are, I think, the only country in which literary papers print weekly lists of best sellers, and these guide the purchases of the lending libraries, the review columns of the newspapers, and most of the reading of the so-called reading public. The only literary paper that attempts broad coverage, *The New York Times Book Review,* provides some token articles and reviews for the cultivated intelligence but for the most part caters to the taste of the millions and the values of the literary market place. The popular author, the purveyor of more or less commercial fiction, writes with one eye on the book clubs and the other on Hollywood. And what books most of our best sellers are, compared with those of a hundred years ago! There was trash then as now, but there were also Dickens and Trollope and Tennyson; our major novelists today very rarely, and poets never, get into the best-seller lists. Or to glance further back, think of our popular songs, the products of Tin Pan Alley, and of the songs of the Elizabethan age or the Middle Ages. Or think of nonliterary figures, statesmen for example, of former centuries: how early many of those men attained maturity of mind, how civilized they were, what style and flavor their speech displayed—and then think of the flat or flatulent utterances of most of our public men. This is not a sentimental jeremiad, it is only a scant reminder of some facts that belief in progress is disposed to overlook.

The Ascendancy of Science

I have spoken of two things that work against the humanities: one negative—the common lack of a desire, capacity, or opportunity for humane self-discipline, for moral and aesthetic cultivation—and one positive—the gross or insidious allurements and pressures of mass civilization. A third is the direct pressure of science and the supreme authority claimed for and given to the scientific method and outlook. Science is of course a large and essential part of a liberal education, and the placing of science in partial opposition to the humanities requires explanation. It is often urged that the scientific imagination and the poetic imagination are very close to each other, if not identical; and doubtless there is a rarefied plane on which that is true—the plane of curiosity, wonder, and the inspired leap of intuition, not to mention fundamental brainwork. But in their normal working these two faculties are very distinct, simply because the natural scientist is largely concerned with physical nature, with things, while the artist is concerned with the nature and behavior of man. Obviously the scientist, qua man, shares the general experience of humanity; and, qua scientist, he may have not only intellectual but moral and aesthetic experience; but it is no less obvious that science per se does not operate in the human realm of moral values and moral choice. Without the humanities and what they represent, science can become merely the instrument of man's aggressive passions.

In the second place, as science and scientific method have more and more come to dominate the modern mind, scientists, philosophers, and laymen more and more have assumed that there is only one kind of truth, one kind of reality—that which is discovered, measured, and verified by science. Viewed in that light, literature sinks to the level of an insignificant recreation or an insidious fantasy of wish-fulfillment. Antagonism or condescension toward literature may not be typical of modern scientists, but such attitudes have many exemplars from Bacon down to Freud and the present moment. A positivist philosopher once put the case to me in a memorable syllogism: The end of life is the contemplation of true propositions; Shakespeare has no true propositions; therefore Shakespeare is not worth reading. So all literature goes into the wastebasket, along with metaphysics and religion.

To register another modern phenomenon from which the humanities especially suffer, it was no doubt inevitable that the immense growth of modern knowledge should lead to subdivision and specialization, but it was no less inevitable that such specialization should be in many ways disastrous. We may remember the old medical doctrine of the four humors: if one of the four grew predominant, the physical and mental constitution became diseased. In the modern world of learning and science such disease is the normal condition. From ancient times up through a good part of the nineteenth century the mass of educated men, including scientists, had a more or less uniform education, a common cultural heritage, a more or less common outlook—however much they might diverge later. But the once seamless robe of Truth has become a thing of holes and lumps and shreds and patches. The fifty departments of a modern university are aggregates of electrons moved by repulsion; and the educated public tends to be an aggregate of heterogeneous groups which do not speak the same language, which share only a minimum of common culture. I am told that mathematicians and physicists, who used to be blood brothers, are now growing apart and unable to communicate with one another. The Modern Language Association meets in a multitude of groups, and the expert in one is an alien in forty others.

It might be said that it is only through narrow specialization that modern knowledge has made such enormous advances. But one may ask if such advances have been worth the price, since the price is the intellectual and spiritual impoverishment of the expert, who becomes little more than a link in an assembly line, a cog in a machine.

I have been speaking of specialization within one's own area of knowledge. The gaps widen greatly of course if we take a larger view of the larger areas—the humanities, the social sciences, and the natural sciences. What do the members of one tribe know of the materials, ideas, and aims of the others? Perhaps I may hazard some guesses; if these are mistaken, no doubt I shall be promptly informed. In the matter of narrowness of knowledge it seems to me that, on the whole, humanists, while not innocent, are somewhat less guilty than sociologists, psychologists, philosophers, and natural scientists. No one can teach or write about literature without taking account of social and economic life, politics, philosophy, and science; but apparently teachers of all these subjects can take small account of literature. It is almost solely the literary scholars who have dealt with the effects on life and literature of all these currents of thought and discovery, from Plato and Aristotle to Marx and Darwin and Freud. And it may not be irrelevant to add what any publisher will tell you—that manuscripts submitted by social scientists and scientists often have to be farmed out for rewriting in decent English.

But mere knowledge of disciplines other than one's own is less important than a sympathetic understanding, and it may be suggested that many scientists and social scientists are so wedded to scientific method and materials that they do not understand what the humanities are and seek to do. European and English scientists are perhaps less open to this criticism than Americans, since they are more likely to have had a humanistic education and are also more likely to continue general reading through their life. We might recall, though, one who did not have a humanistic education, T. H. Huxley. In his debate with Matthew Arnold over the educational claims of literature and science, while Arnold was appealing to man's profound concern with the power of conduct, of intellect and knowledge, of beauty, of social life and manners, Huxley saw "culture" as only belletristic ornament and diversion.

Further, the humanities by nature and definition embrace all kinds of knowledge and experience over the whole of mankind's history, whereas the natural scientist, as such, is concerned only with the latest developments. The history of science is itself an enlightening and important discipline which may well be included under the humanities, but science per se is quite different. The scientific mind instinctively assumes that knowledge means chiefly new knowledge; the humanities, on the other hand, are not measured in terms of novelty and discovery. The humanist's object is to conserve and propagate our cultural heritage, to transmit the wisdom and art of the past to people of the present and future. Naturally there is continual reinterpretation of the great works of literature and art, continual rewriting of the evolution of ideas (and cultural history is a valuable discipline in itself), but considerations of old and new, of time and history, are only secondary to making the assimilation of those great works a richer experience.

However desirous scientists are of enlarging knowledge and improving the lot of man, they can pursue aims and ideas which are remote from the humane. To take a fairly recent and presumably not unrepresentative example of scientific thought in action, *The New York Times Magazine* of 8 December, 1957 had a composite article by six eminent American scientists under the title "Science Looks at Life in 2057 A.D." I cannot summarize the six contributions, but I can quote the puzzled comment by my colleague and friend, Howard Mumford Jones, a humanist who has long shown especially active sympathy with science. Professor Jones wrote thus to the *Times*:

> I read with growing wonder the composite article entitled "Science Looks at Life in 2057 A.D." Nothing so indicates the narrowness of scientific outlook among experts. The only conceivable future is for the regulation of offspring, interplanetary space travel, the control of physical power, the eating of vegetable steaks, artificial photosynthesis, and advanced psychology based upon machines.
>
> That art, that philosophy, that religion, that the whole vast, rich, historical outlook and maturity we commonly associate with the humanities might have some determining part in civilization a hundred years hence seems to occur to none of your distinguished contributors. Man reared from a test-tube baby to machine-conditioned psychology and eating vegetable steaks seems scarcely worth sending to Mars if that is the best he can do with himself.

So said Mr. Jones, with his customary incisive vigor. I submit that no panel of humanists could reveal such stunted and mechanized vision, such an incredible scale of values. If this is the kind of illumination that is to guide mankind through the next century, we may wonder—not about Mars, but if our own anxiety to survive is worth while.

The Age of the Machine

I am not trying to catalogue all the forces and attitudes that tend to blunt and frustrate man's human potentialities, but I should like to touch on some further and familiar characteristics of our time that spring from or go along with the combined pressures of mass civilization and science and technology. These may be lumped under the heading of mechanization or faith in machinery. First, in regard to the literal mechanization of life, we might recall an anecdote about a distinguished museum director who was told that he needed a new machine in his office. After a morning of demonstrations at the I.B.M. headquarters, he, in pensive mood, got into the elevator. During the descent a young woman screeched, because, it appeared, a young man had pinched her. Above the hubbub was heard the voice of the custodian of art: "Thank God some things in America are still done by hand!"

Obviously mechanization has eliminated or lightened much monotonous labor, and has thereby won millions of free hours for old and young. What is done in those millions of free hours is a question one does not contemplate with entire optimism. Like most of our other phenomena, mechanization is not new, except in its present magnitude. At the beginning of the industrial revolution in England Wordsworth remarked:

> For a multitude of causes, unknown to former times, are now acting with a combined force to blunt the discriminating powers of the mind, and, unfitting it for all voluntary exertion, to reduce it to a state of almost savage torpor. The most effective of these causes are the great national events which are daily taking place, and the increasing accumulation of men in cities, where the uniformity of their occupations produces a craving for extraordinary incident, which the rapid communication of intelligence hourly gratifies.

Later, when industrial progress was at its height, Matthew Arnold questioned the sufficiency of a train that took people quickly from an illiberal, dismal life at Islington to an illiberal, dismal life at Camberwell. From those early comments we can jump up to the modern definition of man as an ape in an airplane. For still another kind of mechanization I might quote from Lionel Trilling a sentence I am fond of, a sentence no less grim than funny:

> A specter haunts our culture—it is that people will eventually be unable to say, "They fell in love and married," let alone understand the language of *Romeo and Juliet,* but will as a matter of course say, "Their libidinal impulses being reciprocal, they activated their individual erotic drives and integrated them within the same frame of reference."

One conspicuous symptom of mechanization is blind faith in the group, the desire to belong, to sink one's individual self into colorless conformity without responsibility. This motive operates even in the cultural sphere. Herbert Hoover, when President, anticipated it in the pathetic exhortation: "We are organizing the production of leisure; we need better organization for its consumption." The American belief in free enterprise in the material world has not been attended by an equally fervent belief in free enterprise in the world of the mind. Long before Senator McCarthy was born, to be different from the herd was to invite trouble. The only kind of individuality that is generally admired is skill in sport or smartness in business; the individuality of the cultivated mind and taste meets only indifference or antagonism, in school, in college, and in society. Our democratic religion is the worship of commonness.

Mass Civilization

When we look at American education over the past fifty years or so, secondary education in particular, two combined tendencies are conspicuous—the general pressures of mass civilization and a tradition of anti-intellectualism which is a peculiarly American legacy from Puritanism and the frontier. (Puritanism, of course, had an intellectual tradition too.)

The noble ideal of "high school for all" melted rapidly into a sort of social jelly. Since the majority of the young possessed little or no capacity for intellectual knowledge and ideas, yet must be kept in school, something had to give, and what gave was education. The substantial subjects of study were either eliminated or watered down to the level on which the average and subaverage mind could go through the motions of learning. Then what sheer pressure of numbers seemed to demand as a necessity got strong theoretical support from the multiplying educationists who were translating the doctrines of John Dewey into practice. These doctrines, at least as commonly interpreted, were highly congenial to the numerous Americans who were instinctively and invincibly pragmatic, utilitarian, and anti-intellectual. For Dewey-eyed educationists no one subject of study was better than another, all kinds of experience were equally valuable, doing could replace knowing and thinking; the great end of education was adjustment to the contemporary scene. Thus the humanities, under the opprobrious label of "leisure-class studies," had to be removed or denatured. Foreign languages were useless frills, and of late years more than half of our public high schools have offered no foreign language at all; this cultural parochialism developed during the decades in which the United States engaged in three foreign wars and became the leader of the free world. History, bits of English and American literature, even arithmetic, tended to become branches of civics. The ideal high school has been to a large degree a social service station. The elimination or drastic reduction of intellectual material and of intellectual effort has been carried on in the name of democracy and character-building. These have been nourished by a system which demonstrates that something is to be had for nothing; that the American way of life means shallowness, shoddiness, and the art of "getting by"; and that intellectual and aesthetic cultivation is a mark of pernicious snobbery. In the main, our public schools, instead of opposing mass civilization, have eagerly surrendered to it and fostered it.

To give just one example of the way the educationist mind commonly works, a while ago the poet Randall Jarrell contrasted the pieces of great literature in old school readers with the guff specially written by nonentities for some modern ones. The article drew from the assistant superintendent of schools in St. Paul a protest very typical of our time and country. The idea that boys and girls should know some great poetry and prose of the past belonged to what this man of authority called "an education designed for young aristocrats in a feudal world." One must infer that democracy is founded on barbarism. It was this same educationist who said, in an article defending high schools in reply to the charge concerning foreign languages, that it would take a generation to educate the necessary new teachers. One cannot help asking why, and what the current generations of teachers were taught. Of course we know the answer. The University of California has, under Physics, six pages of courses; under English, six and a half pages; under Education, twelve pages. The Ohio State University offers, under Physics and Astronomy, six and a half pages; under English, the same; under Education, twenty-two pages.

As Robert Ulich (a professor of education) has said, "If in the European countries the time and learning capacity of the able student between twelve and eighteen were as little used as in the United States, they could not survive, nor would their institutions of higher education be what they are." You cannot have many years of very deficient schooling without corresponding effects in the colleges, and we have had them. The state universities have been compelled to take in hordes of unqualified high-school graduates and to spend at least the first two years in trying to do what should have been done not merely in high school but in grammar school. Even the private colleges have to teach freshmen the rudiments of grammar, diction, and idiom. And many of our college graduates, including a number who go on to professional schools, are more or less illiterate.

The Renaissance ideal was not a debased diploma for all, but an aristocratic education for a great many; thus Shakespeare, at the grammar school in the small town of Stratford, was better educated than most of our college graduates. Devereux Josephs, chairman of the President's Committee on Education Beyond the High School, has laid down the principle that "this country will never tolerate the nurturing of an educational elite." As the *Boston Herald* said in an editorial on Mr. Josephs' pronouncement, it is a plain fact of history that "an intellectual elite formed the concept of democracy, kept it alive, and is, moreover, responsible for nearly every progression we have made in science, art, and human relations." And how can you have an intellectual elite without an educational elite?

During the last six or seven years there have been many signs of a mounting revolt—of which Professor Bestor has been a notable leader—among laymen, parents, schoolteachers, and college teachers, a growing demand that slackness and softness give place to serious and substantial education. Then in the fall of 1957 the whole country went into a tailspin over the sudden proof of

Russian scientific prowess. There was a swift and agonized reappraisal of the American school; what we had so continually been told was the wonder and admiration of the modern world now revealed itself as organized babysitting. Displaying our usual emotional instability, we had an immediate wave of zeal for the despised egghead, for "crash programs" in science. It would all be rather comic if it were not tragic. The first large specific consequence was the government's program for financial aid for scientific education. This was not based on any concern for science but only on fear of Russia; astrology or alchemy would have got the same support if they could have helped in the arms race. So, in the middle of the twentieth century, the chief end of American education is the training of military engineers, and our nearest approach to the angels is by way of missiles and spaceships.

What the Humanities Can Do for Man

In sketching the multitudinous things that work against the humanities, I have partly indicated what the humanities have always been and what, if given a chance, they can always be, even if they lack the pragmatic or spectacular appeal of the social or the natural sciences. I assume of course that all three branches of study make unique contributions and are essential elements in a liberal education. But, as I have said, it is clear that many people, including not a few scientists, social scientists, and philosophers, look on the humanities as an outmoded luxury. On the contrary, the humanities are a prime and practical necessity if men and women are to be fully human; they are in fact *the* prime necessity, since they are part of our everyday living and being. But it is not front-page news if a rocket goes off in the mind of John or Mary Smith, if he or she, through absorbing a poem, becomes a person of richer imaginative and moral insight, of finer wisdom and discrimination and stability. For the experience of literature and art is an individual experience, and, like all other really important things, it cannot be measured.

In the educational creed of the Renaissance, the humanities embraced chiefly the literature, philosophy, and history of Greece and Rome. These were the liberal studies worthy of a free man; they dealt with and appealed to man as man, as an intellectual and moral being, not as a professional and technical expert. For many humanists these studies were complementary to divinity, which was the keystone of the arch; for others they remained secular. In the sixteenth century science and mathematics were not yet important enough—except for a few humanists like Rabelais—to become a standard part of the curriculum. To jump up to our time, divinity went overboard long ago; the classics engage a minute fraction of the young or old; history has become a social science; and modern philosophy seems to be suspended somewhere between linguistics and mathematics. In common usage, the humanities are now narrowed down to literature and the fine arts.

The old classical education is usually condemned by its foes because it could suffer at times from low aims and poor teaching and because, so they say, it was narrow and remote. Well, low aims and poor teaching can afflict the most modern studies. As for narrowness, the great virtue of the old classical education was, and is, its combined unity and variety. One body of material, the expression of two great and related civilizations, contained an array of masterpieces in prose and verse which not only gave full play to the imaginative, emotional, aesthetic, and critical faculties, but embraced the first principles of ethics, metaphysics, history, politics, and economics. As everyone knows, this body of material provided in Europe for many centuries—and to some degree still provides—a stable community and continuity of ideas and values. For a long time it governed American education; it nourished the men who wrote the Constitution and the Bill of Rights. The large gap left by the dropping of the classics has been recognized even by educators of lukewarm sympathy, and the vacuum has been filled, after a fashion, by courses in "general education" and the like.

It may be said that a sane and rational approach to the problems of man and society can be achieved through means other than the humanities. Doubtless it can, up to a point—though I think natural scientists as well as humanists may feel depressed when they scan many of the courses offered by our departments of sociology; and I myself, I might say, have little faith in the wisdom of much current psychology, which in some forms seems a crude affront to the dignity of man. At any rate, one of the distinctive things about the kind of illumination to be gained from the humanities is that one is never allowed to forget the individual person, to lose sight of oneself and others in a large blur of social and economic forces and formulas. The basic question, which contains and dwarfs all others, is whether John Smith—or Ivan Ivanovitch—feels and thinks and acts rightly or wrongly, whether he loves what the best human experience has found lovable and enlightening and sustaining. The materials of the humanities are the products of great individual minds, and they work directly upon individuals. As Whitehead declared, "Moral education is impossible apart from the habitual vision of greatness." I do not know where the habitual vision of greatness is found except in the humanities.

From antiquity until recent times, literature has been the chief medium of education, and in spite of—or because of—the emergence of new kinds of knowledge, its responsibilities have not shrunk. They have indeed grown more important than ever, as many things have combined to lower the dignity, benumb the sensibility, and drown the voice of individual man. The aim of the humanities is not to adjust people to life, to the pressures and low ideals of mass civilization, but to enlighten and disturb them, to inspire and strengthen them to adjust life and themselves to the great traditional ideals of the best minds, the saving remnant of the human race. The subject matter of literature is the whole range and texture of life, the material conditions of existence, all that man does and is and would be as an individual being, his desires and hopes and joys and fears and sufferings and defeats and victories; and, along with that, all that he experiences as one of a family, of a community, of a nation, of the human race. Such headings apply to man and literature in all ages, and literature at once reflects, opposes, and creates the spirit of every particular age. Modern literature, the literature of the age of anxiety, has been preoccupied with man's increasing consciousness of his loss of outward and inward wholeness and order, with his sense of being a fragment in a fragmentary world. And the eternal quest of literature, intensified by the common modern loss of religious assurance, has been the quest of order. From the chaos of life and society the artist seeks a pattern of explanation, a pattern that helps him and his readers to achieve some measure of dominion over experience. The value and the success of such efforts depend upon many things, from his own and his readers' powers of imagination to the feeling for structure and tone and the texture of word and rhythm. Historical and philosophical and every other kind of knowledge, including that of everyday life, enter into the creation of and the response to the work of art, but the work in itself is unique, and no discursive substitute can take its place.

Great Writers as the Conscience of Mankind

Since divinity is gone and philosophy has largely abandoned its traditional concern with the good life, it is through literature that most young people now get their only or their chief understanding of man's moral and religious quest. And there has been some change in the spiritual climate over the last thirty years or so. In the 1920's, the time of both brash optimism and cynical defeatism, it would have been hard to find students reading Dante and George Herbert and Milton and Hopkins and Eliot with the real sympathy that many now feel. For the more intelligent and sensitive students of today—a small minority, to be sure—are a serious and mainly a conservative lot. They not only live in our unlovely world, they have no personal experience of any other. They are aware of hollowness and confusion all around them, and of what is still more real, hollowness and confusion in themselves. And in literature they find, as countless people have found before them, that their problems are not new, that earlier generations have been lost also; they may discover, too, that writers of an earlier day had answers which are not out of date but richly worth pondering. Such young people find in literature, literature of the remote past as well as of the present, what they cannot find in textbooks of psychology and sociology and science—the vision of human experience achieved by a great spirit and bodied forth by a great artist. In the humanities, ethical and aesthetic values are inseparably bound up together. The great instrument of moral good, said Shelley, is the imagination; and poetry ministers to the effect by acting upon the cause.

The great artist's vision of life is not dimmed or deadened by time, because it is a distillation of man's finest insights, of his supreme awareness both of what man is and of what man might be. Thus the literature of the centuries of ancient paganism, of the centuries of Christianity, and of our own era, the era it has become fashionable to call post-Christian—all this constitutes a simultaneous and living whole. Whatever the vagaries of Homer's gods, his ethics have an unfailing soundness and rightness. So likewise do Shakespeare's, though Shakespeare's religious creed is very different from Homer's. Shakespeare's highroad leading nowhere, as Alfred Harbage has said, is the road home; what he tells us is what we have always known—though he tells us much more too.

When the slayer of Hector and the father of Hector meet, brought together by the command of Zeus, they both learn the meaning of compassion. The same lesson is learned, through suffering, by King Lear, and he dies, like the aged Oedipus, in the knowledge of love given and received. The mind of Hamlet swarms with ideas and feelings unknown to Orestes, but there are affinities between them. And the religious integrity that unites Antigone with Jeanie Deans bridges the gulf between the laws that grow not old and the God of Scottish Calvinism. The shock of a young man's initiation into the adult world of evil links Sophocles' Neoptolemus with the central figure in Hemingway's *The Killers*. Thus the great writers—pagan, Christian, agnostic—"are folded in a single party"; they are the imagination and the conscience of mankind. If we have any hope of going forward, or of not slipping further backward, they must be our guides.

ERNEST NAGEL

The place of science in a program of liberal education is not a new problem; it has been at the focus of reflection on educational philosophy and practice since Plato. Nevertheless, although the basic issues may not have undergone radical transformation with the passage of the centuries, the problem has acquired new dimensions and fresh complexities in contemporary American society.

Until comparatively recent times, the theoretical sciences were regarded as branches of philosophical inquiry, having for their ultimate objective knowledge of man's supreme good in the light of his place in the universe; education designed for developing enlightened and cultivated minds was reserved for small minorities in relatively small populations; and the organization of human life did not require large groups of highly trained scientific personnel. Under such circumstances, it was easy enough for Plato and his successors to argue persuasively for a conception of liberal education in which the study of science occupied a prominent place. But these circumstances are no longer present. It is in consequence more difficult today to win effective general agreement that a solid grounding in natural and social science is an indispensable part of a humanistically oriented education, and that such grounding is no less essential for the formation of a liberal intelligence than is thorough exposure to the materials traditionally classified as belonging to the humanities.

What are the distinctive contributions the study of science can make toward realizing the objectives of a liberal education? Let me outline what I regard as the three cardinal contributions.

The Theoretical and Moral Value of Science

It has been the perennial aim of theoretical science to make the world intelligible by disclosing fixed patterns of regularity and orders of dependence in events. This aim may never be fully realized. But it has been partly realized in the scientific exploration of both animate and inanimate subject matter. The knowledge that is thus progressively achieved—of general truths about various sectors in nature as well as of particular processes and events in them—is intrinsically delightful to many minds. In any event, the quest for such knowledge is an expression of a basic impulse of human nature, and it represents a distinctive variety of human experience. It is a history of magnificent victories as well as of tragic defeats for human intelligence in its endless war against native ignorance, childish superstitions, and baseless fears. If to be a humanist is to respond perceptively to all dimensions of man's life, an informed study of the findings and of the development of science must surely be an integral part of a humanistic education.

There is the further point that knowledge acquired by scientific inquiry is indispensable for a responsible assessment of moral ideals and for a rational ordering of human life. Ideals and values are not self-certifying; they are not established as valid by appeals to dogmatic authority, to intuitions of moral imperatives, or to undisciplined preference. Moral ideals must be congruous with the needs and capacities of human beings, both as biological individuals and as historically conditioned members of cultural groups, if those ideals are to serve as satisfactory guides to a rich and satisfying human life. The adequacy of proposed moral norms must therefore be evaluated on the basis of reliable knowledge acquired through controlled scientific inquiry. It is simply grotesque to imagine that anyone today can exercise genuine wisdom in human affairs without some mastery of the relevant conclusions of natural and social science.

I am not unaware that there have been great moral seers who possessed little if any scientific knowledge of the world or of man, and who nevertheless spoke with understanding about the paths of human virtue. However, though such men may have expressed profound insights into the ways of the human heart, merely to proclaim an insight does not establish its wisdom; and it is by no means self-evident that their vision of the human good, though generous and wise for their time, is really adequate for men living in different climes and with different opportunities for developing their powers. Insight and imagination are undoubtedly necessary conditions for moral wisdom, but they are not sufficient. For insights and visions may differ, and knowledge of the world and human circumstance must be introduced for adjudicating between conflicting moral ideas. It would be absurd to deny the exquisite perceptions and the stimuli to reflection that are often found in the pronouncements of scientifically untutored moral seers. But I do not believe their pronouncements can be taken at face value, or that in the light of the scientific knowledge we now possess those pronouncements are invariably sound. In short, apart from the intellectual joys accompanying the enlarged understanding of the world that scientific knowledge may bring, such knowledge is indispensable if the ideals and the conduct we adopt are to be based neither on illusion nor on uninformed parochial preferences. It is not an exaggeration to claim that the theoretical understanding that the sciences provide is the foundation for a liberal civilization and a humane culture.

Science as Intellectual Method

The conclusions of science are the products of an intellectual method, and in general they cannot be properly understood or evaluated without an adequate grasp of the logic of scientific inquiry. I am not maintaining, of course, that there are fixed rules for devising experiments or making theoretical discoveries. There are no such rules; and it is in large measure because it is commonly supposed that there are, that scientific inquiry is frequently believed to be a routine grubbing for facts, and unlike literature and the arts to require no powers of creative imagination. Indeed, science has fallen into understandable though undeserved disrepute among many humanistic thinkers because students of human affairs have sometimes permitted this misconception to control their inquiries and their literary productions. Nor am I asserting that the sciences share a common set of techniques of inquiry, so that disciplines not employing those techniques are not properly scientific. Except for the ability to use a language, it is doubtful whether there is such a set of common techniques. Certainly the techniques required for making astronomical observations are different from those used in the study of cellular division; mathematically formulated laws are relatively recent developments in chemistry, biology, and the social sciences; and though quantitative distinctions are widely used in many sciences, the techniques of measurement are often quite different for different subject matters.

On the other hand, I am suggesting that what is distinctive of all science, not merely of natural science such as physics, and what assures the general reliability of scientific findings, is the use of a *common intellectual method* for assessing the weight of the available evidence for a proposed solution of a problem, and for accepting or rejecting a tentative conclusion of an inquiry. Scientific method, in my use of this phrase, is a procedure of applying logical canons for *testing* claims to knowledge.

Those logical canons have been adopted neither as arbitrary conventions, nor because there are no conceivable alternatives to them, nor because they can be established by appeals to self-evidence. They are themselves the distilled residue of a long series of attempts to win reliable knowledge, and they may be modified and improved in the course of further inquiries. They owe their authority to the fact that conclusions obtained in accordance with their requirements have agreed better with data of observation, and have in the main withstood further critical testing more successfully, than have conclusions obtained in other ways. The use of scientific method does not guarantee the truth of every conclusion reached by that method. But scientific method does give rational assurance that conclusions conforming to its canons are more likely to approximate the truth than beliefs held on other grounds. To accept the conclusions of science without a thorough familiarity with its method of warranting them is to remain ignorant of the critical spirit that is the life of science. Not every claim to knowledge is a valid claim; and without a clear grasp of the standards that evidence for a conclusion must meet, the risk is large of becoming a slave to every rhetorical appeal, to every plausible though specious argument, and to every intellectual fashion.

A firm grasp of the logical grounds upon which the sciences rest their conclusions serves to show that the sciences can make no dogmatic claims for the finality of their findings; that their procedure nevertheless provides for the progressive corrections of their cognitive claims; that they can achieve reliable knowledge even though they are fallible; and that however impressive the achievements of science have been in giving us intellectual mastery over many segments of existence, we cannot justifiably assume that we have exhaustively surveyed the variety and the depths of nature. The critical temper, the confidently constructive rationality, and the manly intellectual humility that are essential for the practice of scientific method are not simply adornments of a well-balanced mind; they are of its essence.

Science as the Code of a Community

This brings me to the final point I want to make in this context. Viewed in broad perspective, science is an enterprise carried on by a self-governing community of inquirers who conduct themselves in accordance with an unwritten but binding code. Each member of this republic has the right and the obligation to make the most of his capacities for original and inventive research, to make full use of his powers of imagination and insight, to be independent in his analyses and assessments, and to dissent from the views of others if in his judgment the evidence requires him to do so. In return for this he must submit his own investigations to examination by his scientific peers, and he must be prepared to defend his claims by reasoned argument against all competent critics, even if he should believe himself their superior in knowledge and insight. Accordingly, no question of fact or theory is in principle finally closed. The career of science is a continuing free exchange of ideas, and its enduring intellectual products are in the end the fruits of a refining process of mutual criticism. This does not mean that individual scientists do not possess passions and vanities, which are often obtacles to dispassionate judgment and which may hamper the advance of knowledge. It does mean that the institution of science provides a mechanism for discovering the truth irrespective of personal idiosyncrasies, but without curtailing the rights of its members to develop freely their own insights and to dissent from accepted beliefs.

The organization of science as a community of free, tolerant, yet alertly critical inquirers embodies in remarkable measure the ideals of liberal civilization. The discipline that fosters these qualities of mind therefore must have an important place in an educational program designed to develop members for such a society.

I must now discuss some of the obstacles in modern American society that stand in the way of adequate realization of the values that are obtainable from training in science.

The Danger of the Mole's-Eye View

In the first place, there is the high degree of specialization now required for exploration in most branches of science. Much of the indispensable day-by-day work of the scientist, even when he engages in fundamental research, has in consequence a relatively narrow scope. It is work which for the most part can be carried on successfully without thinking about the basic assumptions and issues. Accordingly, most scientists, whether in academic life or in one of the engineering professions, have at best only a perfunctory interest in the philosophical aspects of their discipline.

To be sure, in the course of solving their own technical problems many of the creative minds in science have felt themselves compelled to give close attention to the structure of scientific ideas, to examine the significance of scientific statements, or to analyze the logic of scientific inquiry. Indeed, revolutionary advances in scientific theory sometimes have been the consequences of just such comprehensive reflections. But to a large fraction of practicing scientists, concern with such matters is a luxury for which their immersion in detailed technical problems leaves them little time.

In consequence, the intellectual climate in which the sciences are taught is not generally favorable to the study of science as part of a liberal education. Courses in so-called "general science," designed in the main for those not contemplating a scientific career, are frequently so empty of scientific content, as well as of competent philosophical commentary, that they are viewed with merited contempt by scientists and humanists alike. On the other hand, the more traditional courses in science are normally organized so as to provide training mainly in professional skills. Unless he is guided by an exceptional teacher, it is only an exceptional student who acquires from such courses a just appreciation of the structure of scientific ideas, or of the logic and the civilizing significance of scientific procedure.

A second reason for the current difficulty in developing a satisfactory program of liberal education is rooted in the fact that in the popular, as in the Marxist, image of science, even fundamental research is primarily a handmaiden to technology. The major financial rewards and positions of influence go for the most part to the applied scientist, such as the engineer and the physician, rather than to those engaged in basic research. Under such circumstances it is by no means easy to secure adequate recognition for the conception that the pursuit of pure science is one of the glories of a liberal civilization. I am not suggesting for a moment that the practical values of training in the sciences are of no serious importance and should be ignored in considering the place of science in liberal education; and I have never encountered a good reason for maintaining that there is no room in a humanistically oriented education for anything in the curriculum of scientific studies that will prepare students for some practical profession. I am urging, however, that primary or exclusive emphasis upon the development of specialized skills is a disservice to the student, to the future of both pure and applied science, and to the prospects of a liberal society.

The Fear of Science as Inhumane

While science is currently prized for contributions to technology, paradoxically it is also condemned as the ultimate source of many of our major evils. The invention of the terrifying instruments of mass destruction has evoked widespread distrust of science, and has reinforced deep-seated doubts whether the benefits of scientific progress outweigh the miseries and the fears that apparently must be included in the price for advances in knowledge. Further, many critics of contemporary society attack science for having made possible the rise of so-called "mass culture" and its alleged consequences: the severe restrictions upon individual freedoms, the loss of a sense of individual inner purpose, the use of techniques for manipulating the minds as well as the bodies of men, the decline in standards of human excellence, and the general vulgarization of the quality of human life. Moreover, it is argued that science is inherently amoral and therefore likely to develop in the student a trained incapacity for distinguishing between good and evil and a callous indifference to humane values. In short, many humanistic thinkers view with concern the assignment to science of any large role in liberal education.

I shall try to meet some of the strictures made in the name of humanism against science and its influence; and to suggest what could be done to emphasize the humanistic and philosophic import of science, without depriving science instruction of substantive content.

It is undoubtedly true that the existence of mass cultures depends upon the technical fruits of theoretical and engineering research. It is also undoubtedly the case that many of these techniques have been put to reprehensible uses. On the other hand, it is absurdly unilluminating to make science therefore responsible for the failings of contemporary society, as unilluminating as it would be to place the blame for Hitler's moral inadequacies upon the procreative act of his parents. Scientific discoveries and inventions indeed have created opportunities which frequently have been misused, whether by design or by inadvertence. But an opportunity does not determine the use that men make of it. It is childish to bewail the expansion of science as the chief source of our current evils, and sheer sentimentality to look with longing to earlier days when science played a less conspicuous role in the human economy—as if living under such earlier conditions were an option now open to us, and as if societies less complex than ours exhibit no failings comparable with those of our own.

It is not possible to deny that, despite improvements

in the material conditions of life for an increasing fraction of the populace, much of our energy is directed toward the realization of shoddy ideals, and that relatively few men lead lives of creative self-fulfillment and high satisfaction. It is difficult to demur at such indictments without appearing to act the part of a Philistine. Nevertheless, the failings noted are not unique to our own culture. Critics of American mass culture tend to forget that only comparatively small elite groups in the great civilizations of the past were privileged to share in the high achievements of those cultures, and that even those groups had only limited opportunities for appreciating the supreme products of the human spirit. In our own society, on the other hand, modern science and technology have made available to unprecedented numbers the major resources of the great literature and of the arts of the past and present, never accessible before in such variety even to the highly privileged and cultivated members of earlier societies. I do not claim that these benefits are of passionate interest to the great majority. But there seems to me ample evidence that an increasing number in our society has come to value them, and that as a consequence of exposure to such things tastes have become more discriminating and less provincial.

Discriminating tastes cannot be formed overnight. In view of the size and the heterogeneous character of the American population, and of the fact that adequate leisure and training for developing and pursuing rational ideals is a fairly recent acquisition for most of its members, it is perhaps remarkable how rapid has been the growth of sensitivity in our society to the great works of literary, scientific, and artistic imagination. It is simply not the case that the mechanisms of our alleged mass culture are all geared to enforcing meretricious standards of excellence, or that there is today a decreasing number of opportunities for men to cultivate their individual talents. The evidence seems to me overwhelming that the growth of scientific intelligence has helped to bring about not only improvements in the material circumstances of life, but also an enhancement in its quality.

Humanists Not Immune to Provincialism

If thorough exposure to the discipline of science is essential for the development of a liberal intelligence, familiarity with the subjects traditionally classified as the humanities is no less essential. The values implicit in the study of the humanities are too well known to require extended comment. Their study acquaints us with a range of human aspirations and passions to which we can be strangers only if we remain provincial members of the human race; they transmit to us visions of human excellence that have stirred men throughout the centuries and that make men kindred spirits despite accidents of birth and circumstance; and they make us conscious of our cultural heritage, and so potentially more discriminatingly aware of its virtues and limitations. But there is no inherent incompatibility between the liberal values implicit in the study of the sciences and those fostered by the humanities.

Nevertheless, professional humanists often display a snobbish traditionalism, a condescension toward everything modern, and an impatience with the critical standards of scientific thought. There have been humanists whose enthusiasm for the aqueducts of ancient Rome had no bounds but for whom contemporary systems of water supply were undeserving of a cultivated man's serious interest. There are self-proclaimed humanists who are profoundly affected by the tragic heroism of the Spartans at Thermopylae but who dismiss the Warsaw uprising against the Nazi oppressor as merely a sordid incident. There are humanists who claim a special kind of truth for knowledge about human values and who reject as sheer presumption the view that ordinary canons of scientific validity are pertinent for assessing the worth of moral insights. Needless to say, I am not suggesting that attachment to the classical tradition of humanistic thought is invariably associated with the narrowness of spirit some of these attitudes exhibit. However, the fact that they are sometimes manifested suffices to show that professional scientists have no monopoly on snobbery and provincialism, and that training in the humanities does not insure breadth of perspective.

Since the claim that the humanities represent a distinctive mode of knowledge is a frequent source of antagonism between scientifically oriented thinkers and spokesmen for the humanities, I must deal with it briefly. The claim seems to me to rest partly on a misunderstanding of what is covered by the label "scientific method," partly on a confusion of knowledge with other forms of experience, and partly on what I regard as a mistaken belief in the efficacy of human reason to establish absolutely certain and necessary truths about empirical subject matter. I have already indicated that the label "scientific method" does not signify either a set of rules for making discoveries or the use of certain techniques in conducting inquiries. In any case, I am in full agreement with those who maintain that distinctive subject matters require distinct modes of investigation; that the techniques employed in the natural sciences are not paradigmatic for the study of human affairs; and that though there are physical, biological, and sociopsychological conditions for the occurrence of preferences and valuations, trustworthy judgments about moral ideals cannot be deduced simply from statements about those conditions.

The disciplines constituting the humanities in some cases supply instruction which is no different in kind, though it is different in specific content, from the knowledge obtained in various special areas of natural science. This is patently the case for history and biography, and for much descriptive literature about the habits, customs, and aspirations of men. The factual claims of such literature must be tested by reference to the available evidence. Though standards of proof in these inquiries may be less stringent than in other areas, cognitive claims are validated through the use of logical canons common to all discursive thought about empirical subject matter. On the other hand, there are other humanistic disciplines, among them poetry, painting, and music, which are sometimes alleged to be sources of a special kind of knowledge, to which the canons of scientific method are said to be irrelevant. Now undoubtedly these disciplines can be instructive, in ways different from the way propositions are instructive. They can provide objects for reflection and perception; they can acquaint us with works of imagination that develop our sensibilities and heighten our powers of discrimination; they can present us with patterns of sound, color, and rhythm which evoke, intensify, and discipline emotional responses; and they can confront us with embodied visions of human virtue and human destiny. These are all important and instructive forms of experience. But since nothing is stated by these objects in propositional form, in no intelligible sense can they be regarded as conveying truth or falsity. They are therefore not sources of a special kind of knowledge, though they may be occasions or subject matters for knowledge.

Men who are equipped by native endowment and training to be successful investigators in one area are usually not equally successful in other areas. In any event, the capacity for making contributions to moral enlightenment is not uniformly distributed; and insofar as humanists are capable conservers and purifiers of the conscience of mankind, they require powers of moral imagination and insight which are as specialized as are the powers of imagination and insight into physical processes that the competent physicist must possess. However, as has already been argued, imagination and insight are not sufficient for establishing a cognitive claim, whether in morals or in physics. For insights must be tested. In a manner analogous to the procedure in physics, a test of a moral insight consists in formulating a hypothesis, comparing the consequences of the hypothesis with alternative assumptions and with empirical data relevant to the problem, and finally evaluating the adequacy of the hypothesis as a solution of the moral problem in the light of the evidence. Those who reject this procedure as not suitable for the adjudication of moral issues, and who also reject authoritarian justifications of moral judgments, attempt to validate moral principles by appealing to an alleged rational intuition of their necessary truth. I do not think this approach is tenable, among other reasons because of the historical fact that men have claimed intuitive certainty for incompatible moral principles. Accordingly, though there are as many distinct true statements as there are situations about which predications can be made, there are not several *kinds* of truth, and there is only one reliable *method* for establishing claims to truth. In short, the contention that the humanities employ a distinctive conception of truth and represent a mode of knowledge different from scientific knowledge seems to me to be the consequence of a failure in analysis.

I have already indicated what I take to be the humanistic values fostered by science. Finally, I want to suggest briefly how some of these values might be conveyed through the teaching of scientific subjects. The prospects for a liberal society depend upon the teaching of science as part of a liberal education which is dominated neither by a narrow utilitarianism nor by a comparably myopic professionalism. We cannot afford the folly of killing the goose that lays the golden eggs. Whatever be the urgencies of foreign or domestic politics, we must not institute a system of science education whose primary aim is to prepare students for careers in applied science. We must aim also to develop capacities which will contribute to the pursuit of the disinterested love of learning along the entire front. But above all, somewhere in their education we must equip the future laymen in science, as well as the future engineers and the future pure scientists, with mature conceptions concerning the nature of the scientific enterprise and the logic of scientific inquiry.

My main suggestion is that the liberal values of science can be best exhibited by teaching science with a strong emphasis upon methodological issues. I most certainly do not mean by this the institution of courses in the methodology of science, separate from or in lieu of work in the substantive materials of the sciences. Such courses, though they have a place in the curriculum, convey little to those unfamiliar with the subject matter of the sciences. I do mean that the materials of a discipline should be so presented that the logical principles controlling the analysis, the organization, the validation, and the modification of scientific statements are kept in full view, and that the findings of the sciences are exhibited as the products of a creative but critical intelligence.

The Importance of Method

The student should be disabused of the common misconception that a collection of facts is either the beginning or the goal of scientific inquiry. Emphasis must be placed on the theoretical motivations that underlie the gathering of data, upon the selective character of observation and experiment, and upon the need to analyze and interpret the primary data of observation before they can be admitted as significant fact. Moreover, the student should be made to recognize that the concepts to which he is introduced have not been obtained by a process of simple abstraction from empirical data, but that they are intellectual creations, often *suggested* by the data, and are the products of a constructive imagination. Accordingly, the structures and functions of scientific concepts need special attention, and the logical as well as overt operations that relate concepts to crude experience must be discussed. In view of the increasing role that quantitative notions play in modern natural and social sciences, the logic of measurement and the major types of quantitative measures occurring in a given discipline deserve particular consideration. The student should be made aware that the introduction of quantitative distinctions is not a denial of qualitative differences, but on the contrary is a means for identifying such differences in a more discriminating manner than is customary in everyday affairs. The chief objective of these methodological emphases is to make explicit that science is not a mechanical routine, and that even at the most elementary levels of achievement it involves the use of a disciplined but sophisticated imagination.

The Function of Theory in Science

A second group of considerations to which attention needs to be given is the various ways in which the materials of a subject matter are organized and explained. The student should come to understand that natural history is not natural science, and he should be taught to appreciate the difference in this respect between, say, classificatory botany and genetics. Here again the realization should be enforced that theories are not extracted from empirical data, that they are not inductive generalizations or extrapolations from the observed facts, that they are indeed free creations of the mind. On the other hand, it should be made clear that theories are not just arbitrary inventions, and that they must meet a variety of conditions to be satisfactory. Accordingly, attention needs to be given to the function of theories both as guiding principles for further inquiry and as unified systems of explanation and prediction. It is essential in this connection to note the limitations of a crude empiricism and to stress the intellectual and practical advantages that follow from theories which enable us to subsume under a few principles a vast array of apparently disparate facts. Furthermore, the student should be enlightened on what really takes place when a theory, initially adopted to account for a limited range of similar phenomena, eventually absorbs into its scope of application quite different phenomena. Such extensions of theories occur repeatedly in the history of thought, and they effect important unifications in our knowledge. But such unifications are often construed to signify that science has somehow managed to diminish the variety of nature and to destroy apparent distinctions. In consequence, science comes to be conceived as a form of black magic which converts the world into an alien mystery and makes incoherent the procedures of science itself. Surely one thing a student ought not to carry away with him from his exposure to science is the belief that the universe becomes less intelligible the more science advances.

Science as the Discipline of Judgment

Finally—and this seems to me of greatest importance—the substantive materials of every science should be so expounded that the student acquires a habitual sense of the difference between competent and doubtful evidence, and between well-grounded conclusions and those that have a precarious foundation. The basic ideas of the logic of sampling procedure and the rationale of experimental control groups must therefore be brought home to the student. It thus should become clear to him that the mere agreement of a given hypothesis with empirical data does not constitute a sound basis for accepting the hypothesis, and that unless evidence is produced and analyzed with a view to determining what alternative assumptions are compatible with the facts, one has not even begun to think scientifically. It is not an unreasonable conjecture that if these elementary but basic points in the logic of proof were transmitted through the teaching of science, there would not be such a dishearteningly large number of victims to specious claims and preposterous intellectual fads. But however this may be, the student should also become aware that evidence can have different degrees of probative force, depending on the composition and the mode of obtaining the evidence as well as upon the character of the conclusion the evidence is used to support. Accordingly, no one has received an adequate education in science who does not realize not only that science does not claim definitive finality for its conclusions, but also why such claims cannot be made. Those who acquire such a realization also understand the nature of the continuing critical reflection that is essential to scientific inquiry.

Let me add that such methodological emphases in the teaching of science can be achieved only at a price. A large fraction of the price is that such courses cannot be encyclopedic compendia of whatever might be useful for a professional career in science. Some traditional subject matter must therefore be eliminated. I am convinced it can be done without serious loss if the remainder is effectively presented with a stress on those traits that distinguish science as a method of inquiry. But the price is not too high if the teaching of science contributes vitally to a liberal education.

Discussion

BUSH: I certainly do not attack science; that would be absurd. But it is not being hostile to science to say—as not a few wise scientists have said—that in an age dominated by science and technology we need more of the humanities.

Everyone would surely agree with much or most of Professor Nagel's admirable statement on the value and necessity of scientific knowledge and scientific method. I am moved to dissent, however, from the doctrine that all real knowledge, all real truth, is scientific truth arrived at by scientific verification. I would suggest that such a doctrine is itself unverifiable; it is, ultimately, a dogma, which requires an act of faith. No one can know enough about the nature of man and the universe to be able to set up such an absolute.

Moreover, Professor Nagel goes on to contend that moral issues must be resolved "on the basis of reliable knowledge acquired through controlled scientific inquiry." The conclusion is, then, that man the moral being lives, or should live, entirely by the critical reason working scientifically on scientific data. That is surely not so; and one might add that it is not even desirable. The central and dynamic forces and motives in human nature, for good or ill, are at least as much intuitive and emotional as rational. How would scientific inquiry go about evaluating such a profound paradox as "He that findeth his life shall lose it: and he that loseth his life for my sake shall find it"? Does a scientist or scientific philosopher undertake the most important business of his life, falling in love and marrying, by a process of controlled inquiry? In short, to quote a famous mathematician and scientist, the heart has its reasons, which the reason does not know. Or, to quote again the familiar saying of another scientific philosopher, "Moral education is impossible apart from the habitual vision of greatness." How much of that vision depends on scientific verification?

NAGEL: I cannot believe that Dr. Bush and I differ at all on the place and importance of the humanities in education or the life of man. What, then, is the issue? Let me quote an important statement he has made: "The aim of the humanities is not to adjust people to life, to the pressures and low ideals of mass civilization, but to enlighten and disturb them, to inspire them and strengthen them to adjust life and themselves to the great traditional ideals of the best minds, the saving remnant of the human race."

If I understand anything about the history of science, its mission has always been to "enlighten and disturb" people, and to "inspire and strengthen" them. But it is on the point of adjusting to "the great traditional ideals" that I find a principal difference between Dr. Bush and myself. I don't think it is an adequate vision of a liberal education to make us adjust to "the traditional ideals," unless they can be demonstrated to be worthy of being pursued. I have the feeling that Dr. Bush values the emphasis of the humanities upon traditional ideals, whereas my belief is that the validity of those ideals must be scrutinized in the light of whatever evidence we have available; they need to be examined, analyzed, evaluated, judged to be sound or unsound.

Dr. Bush pokes what I suppose is warranted fun at my attempted defense of mass culture. It is difficult to protest against his condemnations without giving the appearance of endorsing the crudities and vulgarities that one finds in our civilization. But I do want to break a lance for two things. In the first place, the developments in science have made accessible to a tremendously large number of people achievements of what I think Dr. Bush would call the best minds of the past. People can, if they will, listen to music which in times past was seldom played even in the private salons of the nobility; they can see great paintings, or at least excellent reproductions of them; they can read a great variety of forms of poetry and prose literature. Science has created these opportunities. So I don't respond sympathetically to this indictment of mass culture. It seems to me it has great potentialities. To ignore those potentialities and say that what we are interested in is the small number who are the saving remnant of the human race is to take an attitude toward mankind that I myself find rather narrow.

The second difference that divides us is the issue whether the humanities provide us with a distinct kind of knowledge. I realize that a good deal of our knowledge comes from what may be called intuition. There is a familiar story to the effect that Schiller kept a rotting apple on his desk, and that its odor inspired him to write great poetry. Now, it may well be that smelling decaying fruit, or going to church, or drinking, or consorting with women of low repute may give a stimulus to intellectual or artistic effort. But you do not decide the excellence of a product by examining the occasion that led a man to the idea. Further, when you claim that something is a worthy human ideal, it has to be evaluated by some sort of test. To say that a man beheld a great vision, as St. Paul did on the road to Damascus, may imply a correct historical claim about the conditions under which many men obtain moral insights. But having that experience itself does not validate the adequacy of the insight. So many men have had intuitions saying so many contradictory things about what is the human good. If you are going simply to fall back upon intuition or authority, you have no way of settling the matter except by setting one dogmatic claim against another. But the method of science provides a way by which the conflicting claims of diverse intuitions can be settled.

I would say of the forms of experience which one calls the humanities—music, painting, or poetry—that if they are not stated in propositional form, I do not know in what sense one can claim validity for them. Are you claiming truth for such experience, truth that you can formulate in propositional form? If you say no, it isn't that kind of truth, but a non-propositional truth, well, I am quite prepared to admit that such experiences may have tremendously inspiring value in human life; but to call them truth seems to me to be debasing language.

Insofar as non-scientific experiences, cherished by the humanities, can be stated in propositional form, their validity must be established in the same way as any other product of discursive reasoning. If, on the other hand, they cannot be stated in propositional form, then they have to be evaluated, not in terms of knowledge, but of what they contribute to the discipline of impulse, perception, and imagination.

LINDSAY: It seems to me there is a great deal of nonsense on both sides in this problem of the relation of science to the humanities. We should emphasize the similarities between the essential methods of science and the essential activities of the people who work in the humanities. Without imagination there could be no science. The intuitive power of the mind in dreaming dreams is the essential basis for the advance of science. I think this has been shown in the great physical theories of our time: the theory of relativity, the quantum theory, and to a certain extent also in the theories of biology. Here we can see the similarity between the actions of the humanist who works with intuition or imagination and the creativeness of science. The creation of a scientific theory, to my mind, is ultimately just as much a work of art as the painting of a great picture or the composition of a piece of music. It represents exactly the same kind of mental activity.

HOOK: Professor Bush quoted Whitehead to the effect that "moral education is impossible apart from the habitual vision of greatness." Then he adds, "I do not know where the habitual vision of greatness is found except in the humanities." Does he mean to imply that the life-work of men like Pasteur, Kepler, Galileo, Bruno, Darwin, and Peirce is altogether free from the qualities of greatness—of courage, of independence, of the ability to stand alone, of patience?

BUSH: I wouldn't, for a moment, deny greatness to scientists. But I was thinking of the kind of greatness that has to do with moral and imaginative construction, not a greatness that attaches to the biography of great men generally.

HOOK: Isn't there something imaginatively great in thinking in such a way that you turn the whole Ptolemaic universe into the Copernican one, and set the world upside down? Isn't there imaginative greatness in thinking of the origin of man, in saying that maybe he was not created by God; maybe he was not an Aristotelian immortal; maybe he developed from other species? Isn't that an exciting idea? Why the prejudice against finding imagination in science?

BUSH: It isn't that kind of greatness that I meant. I meant the kind of greatness that has to do with moral values.

HOOK: Let us take a further question about history as a humanity. Professor Bush, do you know of any other way of testing the truth of an hypothesis in history than by drawing the consequences of the hypothesis and looking for evidence for them? History, you will agree, falls under scientific method?

BUSH: History which is not art.

MORGENTHAU: Well, obviously the ascertainment of historic facts must proceed by scientific method. But the question arises whether history consists of nothing but the verification of historic facts—who did what at what time and under what circumstances? I would regard it as a legitimate historic statement that Lincoln was a greater president than Harrison. And how would you prove that scientifically?

HOOK: Easily, if you tell me what you mean by "greater." You may say that a greater president is a president who has to settle questions of greater difficulty; and you may determine the difficulty of a question by the number of elements which enter into it; and with these things in mind, we might examine the questions which Lincoln decided.

MORGENTHAU: I would never dream of meaning that. When I say that Lincoln was a greater president than Harrison, I have in my mind the image of what a president ought to be, what the ideal type of a president is, and also what is the ideal type of a great man, who fulfills completely the potentialities of a man. And I measure the qualities of a president by those standards which I carry in mind.

NAGEL: Suppose one were to say, this circle is more perfect than that; it seems to me that this is comparable to saying Lincoln was a greater president than Harrison. In each case you have a certain standard and you say one object conforms to that standard to a greater degree than does another. It is a question of fact, once we have agreed as to what the standards are. To be sure, you might offer a criterion of greatness which I would disagree with, but that would be essentially a semantic question, wouldn't it? That is, we could drop the word "greatness" and discuss the standards you are employing.

SHUSTER: I have a little point I would like to raise for the sake of my education. I will present a line of Shakespeare, which is very simple: "Shall I compare thee to a summer's day?" Now if I understand you correctly, your method of proceeding with this poem is, first of all, to ask whether or not the following proposition is tenable: Does this lady suggest a summer's day? Or you can approach it scientifically, and you can say: The adrenalin supply in this gentleman went up at the sight of such-and-such a girl, and consequently what we ought to do is measure the supply to see whether the asserted effect can be verified. From our point of view——

HOOK: Are you serious or joking?

SHUSTER: A little of both. What I want to know is, is the existence of a poem a verifiable proposition?

NAGEL: If you say the poem exists, it is.

SHUSTER: But I mean the thing that makes it a great line of verse. Does that exist?

NAGEL: Shall we distinguish two things: the line, and the statement about the line which says it is a great line of poetry. What you are asking, I take it, is: Can I certify the statement, "This line of poetry is great"? This takes us back again to the question that Dr. Morgenthau raised, whether Lincoln was a greater president than Harrison. And it seems to me again, that before you can decide the question whether this is a great line of poetry, you have to state what you consider to be a great line of poetry, to say such-and-such are the earmarks of a great line—which may require a good deal of exploration. Then, on the basis of seeing whether this line has those qualities, you decide whether it is a great line.

SHUSTER: Then I think you would kill the humanities. This is a great line of poetry because you know it is.

KUSCH: Mr. Shuster knows it is. This is a different thing from "one knows it is."

NAGEL: Of course when I read the line of poetry, I don't stop and say, "This is a great line of poetry, let's enjoy it," any more than I do the like with music or painting or a human being. Your responses are what they are. But if a question is raised, I can settle it only by what I call the scientific method. I don't mean that I am going to use a chemical test or brass instruments in doing this. I am going to proceed in exactly the way I would in the Lincoln case—in terms of some accepted standard of greatness. To be sure, you might simply use the word "greatness" without stopping to analyze it. But that is not a responsible intellectual judgment.

KENNEDY: When you used the analogy of two figures, one being more perfectly circular than the other, I'm afraid you may have slid over the crucial difficulty. It is far easier to apprehend a standard of circularity than a standard of greatness in men or in poetry.

Return to the Curriculum

COHEN: I think that our discussions have established that the humanities and sciences have entirely different methods, deal with different kinds of subject matter, and have different inherent purposes. Assuming that we agree on the need for both the sciences and the humanities, our question is what kinds and phases of the humanities and sciences belong in education, bearing in mind the differing needs of learners at different stages in their development, and the limited potential of those who are to be beneficiaries of education. The ends of education should be defined with sufficient precision to give us a criterion for selecting the disciplines or exercises by which children or young people shall attain these ends.

BUTTS: Could we make this quite specific? It would be useful if the scientists and the humanists would say whether at the high-school level there should be a different approach to the humanities and the sciences for the top 20 per cent of students, the middle 60 per cent, and the lower 20 per cent. Should we water down the offerings in the sciences and the humanities to the 80 per cent who make up the average and sub-average groups?

BESTOR: Since it has been suggested that I believe in an intellectual elite, may I point out that I do not believe in creating one; I do believe that students should be brought along as fast as they can go. I would object to a policy which proposes something basically different for the 80 per cent. Our curriculum should be based on a common schooling for all at their own level. History is as relevant for the 80 per cent as for the 20 per cent; though obviously the secondary schools cannot carry an entire population as far in history, foreign languages, mathematics and science, as they could an elite.

BUTTS: Would there be a different kind of history for the slower students?

BESTOR: A slower-paced one.

BUTTS: Not as intensive?

BESTOR: This would be determined differently in different fields. In mathematics, for example, it would probably be a question of giving the course more slowly, so that in the twelfth grade less gifted students wouldn't have advanced so far. As for history, I would like to think a little more about whether they should have the same, or a more generalized course. But I can hardly think that there are two histories, one for the brighter and one for the less bright.

BUTTS: It is common to have world history in the tenth grade, say, and American history in the eleventh. Would this mean that in world history we should not try to cover the whole range, but hit the high spots, so to speak? And that in American history we should give the abler students more to read and more difficult materials, perhaps getting into original sources, and that the lesser students would have an easier textbook? Would this be what is meant by a slower pace?

BESTOR: I rather imagine it would be that.

BUTTS: That helps me. Could we put the same question about the humanities and the sciences?

BUSH: I agree with what Mr. Bestor said. In the humanities one general trouble has been that not enough has been expected of poor students, that they have been too readily dismissed as not capable of doing anything. I have always had a feeling that poorer students could do more if more were demanded of them, and I think that more should be demanded all along the line—much more, of course, from the best. I think I yield to no one in my admiration for Scott and Dickens; I am one of the few people left who enjoy Scott. But it distresses me to hear, as I do from people from various schools, that when they were in school the books assigned were *Ivanhoe* and the *Tale of Two Cities,* which show Dickens and Scott at their third rate. I remember that Benchley once said he had read one novel in his life, and he didn't know that; his parents moved a great deal when he was in school, so he was always just beginning *Silas Marner*. Of course *Silas Marner* has its advantages as a short work.

I will mention one other thing, of which I have only a very sketchy knowledge. That is the way in which a number of classics have been watered down by the elimination of any words of more than one or two syllables, and by the cutting out of large portions of important matter in order to make them more agreeable. I think students always ought to have things too hard for them; whether they are inferior or high-grade students, they always ought to be stretched. And, speaking now not of very elementary students, but of students in high school, they ought to be reading originals and not things that have been processed for juvenile consumption.

BUTTS: Does this mean that the 80 per cent should read exactly the same things as the 20 per cent?

BUSH: Not so many, perhaps, and less difficult things. Some high-school students might read *The Brothers Karamazov*. But that certainly would not do for the mass.

BOEHM: We would get into trouble with this program because our students are not ready for it. Some people say that if the dullest people in Russia can speak Russian, Americans ought to be able to learn the language, but I wouldn't want to have more than 20 per cent of our pupils undertaking Russian at this time. I think it is a wrong conception that all the children should take everything. If this kind of program were pushed, it would either kill the kids or drive them out of school at the seventh grade.

BUSH: I didn't know I was making a rash proposal in suggesting that more be expected of most students than has been expected of the lower 20 per cent. Surely we do not need to stoop so low to win the students' attention.

BOEHM: I agree. But I don't agree to the contents being the same for the 80 per cent as for the 20 per cent.

BUSH: I didn't mean that, either. The top layer can obviously read more, and more difficult, books.

BOEHM: They can be reading one or two years ahead.

VAN DEN HAAG: I wonder what is the advantage in keeping the bottom 20 per cent in school at all.

BOEHM: You will always have the lower 20 per cent; if you cut them off, you will still have the lower 20 per cent of those that remain. . . . We are very seriously debating in our state whether the attendance law is not too high. But until the Welfare Department, or other parts of Government, know what to do with these boys and girls, we will keep them as wards of the school.

VAN DEN HAAG: For custodial purposes, you mean?

BOEHM: We try to do the best for them we can under the circumstances.

VAN DEN HAAG: You just pointed out that one effect of Mr. Bush's proposal might be to throw these pupils out of school. I want to know what the disadvantage would be.

BOEHM: In our large cities it would be a very serious disadvantage. In New York City and in Philadelphia, as the papers show, the problem of these children of fourteen, fifteen, and sixteen, is a serious problem. If we excluded them, all we would do is to put them out on the streets and have the Government take care of them elsewhere.

VAN DEN HAAG: What you are really saying is that the schools are giving them custodial care.

DIEKHOFF: There is an advantage not only to society in keeping them off the streets, but to the child in giving him the chance to learn something. The primary question is not whether these children are a nuisance to the staff, but whether, as human beings, they deserve an opportunity to learn. My answer to this is "yes."

ALLEN: Under some circumstances you can give them the kind of challenge that Mr. Bush is talking about.

DIEKHOFF: What we have to do is to invent ways to challenge different kinds and levels of minds.

ALLEN: We have assumed that they can learn on their own, and hardly attempted to challenge them.

The Dull and the Gifted in Science

NAGEL: I would agree with Dr. Bush that there is a tendency to make too small demands on children. I have two boys who are neither unusually gifted, I think, nor unusually stupid. They go to a very good school in New York. I don't think that the demands made upon them come up to what they can do, even though this is supposed to be a school where children are pretty highly selected.

On the other hand, I taught mathematics myself in the secondary schools for some three or four years, and I know there is a tremendous variation in ability. And it seems to me just absurd to think that you can give the same kind of material to those who are particularly well equipped and those who are not.

I think you are kidding yourself when you say, "We want a common cultural background for everybody, so we can give everybody the same amount of material." Here I would be inclined to agree that differentiation in the content of science, or at least of mathematics, for those who are well qualified and those who are less able, is necessary. About the positive sciences, such as physics and biology, perhaps Dr. Kusch can inform us.

KUSCH: I would say that it is perfectly possible to teach a college course to a very large group of students, and turn out students to whom you can give A-plus with every confidence that they have acquired a masterly knowledge of the subject. Honest teaching also requires a considerable number of C-minuses. But my impression is that the two kinds of students don't necessarily interfere with one another, at least if they have imaginative teaching. In fact I think it is a poor teacher who cannot stimulate an A-plus student in a class in which the average ones, let me say, are C-plus. On the whole, I am not tremendously impressed with the need of separate classes and separate facilities.

I also have had some experience with the New York system of schools. My oldest daughter goes to public high school. On the college entrance examination board she made a record with which I can say, even with modesty becoming to a father, that I am thoroughly pleased. This was not even a specially selected high school in New York City; it was the Julia Richman Public High School.

SHUSTER: A good school.

KUSCH: Yes; she got a first-rate education. But I am not convinced that these criteria of selection are infallible or have a very high validity. I am not wholly convinced of the merit of a detailed multiple-track system, although I admit that there is a body of downright incompetents who might very well learn something else. If they learn to use a screwdriver, that is all you can expect of them by way of "science."

LINDSAY: Might I put in a word here about the teaching of science on the college level? I wonder if Professor Kusch would recognize the desirability of a different type of science instruction for those students in college who are not intending to enter upon any career that would remotely connect them professionally with science.

KUSCH: Yes! I would wholly support that. The difference in this case is not one of quality; it is essentially one of interest.

LINDSAY: This raises a serious problem which we could argue for a long time. Scientists have by no means reached an agreement as to the way in which science can be presented most effectively to college students who are not going to become professional scientists. The people in the humanities indict us scientists for not facing up to this problem, as they think they have faced up to it in presenting the humanities to the general run of college students. I am inclined to agree with them that the scientists have not done what they should with this problem of science for general education.

KENNEDY: I would like to say that we faced this problem at Amherst and came out with the conclusion that we wanted to teach real science to everyone. We have in our freshman year a combined course in physics and mathematics. Every student is required to take and pass it. They get into calculus by the end of the first semester. At the end of this course they are qualified to go into Physics 22, a difficult course.

I think we have solved the problem in question by teaching science to non-scientists in the same way as to scientists. But this means that we try to teach science in a way that will give the students an insight into the nature of scientific thinking along the lines suggested by Professor Nagel, rather than piano exercises, the mere acquisition of techniques, which they can acquire later in other courses if they intend to go on as science majors.

Furthermore, the interest in science has grown. Before we instituted this program, only 20 per cent of our students were taking a major in science; today a third of them are. We have now got to the point where we almost have to discriminate against some of the students who want to come to Amherst specifically to prepare for scientific professions, because there are so many.

HOOK: How do you select your students?

KENNEDY: That is to the point; it is a highly selected group of students. But any of the institutions mentioned by Professor Nagel, such as Columbia and Princeton, could do the same thing if they wanted to; with a highly selected group of students, it is perfectly possible.

BOEHM: I don't think we can assume that the secondary school can handle science in the same fashion. We are just as unhappy as the colleges are with the present teaching of science to the so-called general education people. Too often the person who is prepared to teach biology and physics teaches his academic science specifically for college preparation, and then the students not headed for college are killed off; they flunk out of these courses early because they are not prepared to handle the type of instruction. We are not doing the right kind of history or literature teaching, either, to these people below average in the secondary school.

Another thing is that we are talking here about the suburban and large city high school. We forget the smaller high schools of a few hundred students or less. There, again, the only way the students get physics is to take academic physics.

TUMIN: Are you advocating the separation of the gifted from the non-gifted? Is there any evidence of improvement through separation?

BOEHM: I am saying that the present teachers in our secondary schools are not prepared to give the diverse kinds of instruction required.

TUMIN: Suppose you had to meet the problem tomorrow. Would you separate high-school students into these different groups?

BOEHM: Physics is already a highly selective subject, earmarked for those who go to college, and if you make it too difficult, they flunk out. We need physics courses which are not an academic preparation for college. We have some physics under the name of "consumer science." This is an attempt to water the subject down—if you must use "water down" for people of different levels of ability. But unless we do that in some measure, the material will be taught as a purely academic subject. The watered-down course reflects the demand in this country at the present time and the teachers we have to work with.

Segregating by Ability

ALLEN: I would segregate students by ability in the major fields; it is certainly worth a trial.

TUMIN: Is there evidence that it works?

WOODRING: I think we have lots of evidence that it works. It is not the kind of evidence that lends itself easily to sharp measurement, but, in the judgment of teachers, students, and their parents, the experiment works. And insofar as you measure the results on achievement tests, it looks very good.

I have visited a number of schools which are using ability groups. One in San Bernardino has been using it for years, and everybody is satisfied with it. The schools that haven't tried it are afraid; they are afraid the parents will object; in one of the suburbs of New York the parents objected. But by and large parents elsewhere have made very little objection. The plan is to divide differently for each subject. So a student may be in the fast group in mathematics, in the slow group in music, and in the middle group in history.

TUMIN: When you say it works, are you basing this on the expressed satisfaction of the middle-range and low-range students, who no longer have to push hard to compete with the brighter students?

WOODRING: No; there is evidence on achievement tests for all three groups. And the bright students are satisfied, because for the first time they are in a group where they can talk about high-level things without being accused of being eggheads by the other students, and where they can approach much more difficult problems and take pleasure in doing it.

BOEHM: A form of separation is already widely in effect. The abler students, moving faster through the grades, are taking the same subjects a year earlier than the students who are average or below. This is the same thing as segregation, if you want to call it such. But it works; it challenges minds on the level on which they should be challenged.

❊ *Philippe LeCorbeiller*

5

The Crisis in Science Education

PHILIPPE LE CORBEILLER

The major problem in American education just now is how to make adults realize that they are living in an age of science. If this problem could be solved, the schools, which are under the direct control of public opinion, would automatically readjust their programs; they would prepare the young for living in their own life span, a life which will command and apply far more science than ours does.

This new "age of science" in which we are living

began around 1900. Since 1900 man has learned to fly, to talk across any distance and to millions of people at the same time, to exploit the energy of the interior of atoms, and to create new moons.

Changes of such magnitude within so short a time are still something of an enigma. We do not yet understand the relation of cultural to biological evolution. Some kind of mutation must have occurred in man around 1900, for surely this "man" who flies, talks across the ether, and uses nuclear energy is not of the subspecies which, one hundred years ago, did none of these things—just as the group of primates who first kept a fire going was something new among the land animals that for three hundred million years had run away from fire.

As a result of radio alone, the whole of mankind has all of a sudden become a single unit, welded by the existence of instantaneous communication. Illiterate masses who, all over the world, learn daily from the village radio what happens in Baghdad or in Little Rock care more for science, because of its technological consequences, than cultured opinion does in the world capitals; for the majority of men, underfed and disease-ridden, are counting on science to give them food and health.

What Is Science?

Science is power: the power of doing things that have never been done before. No one can set limits to what science might achieve. This was the meaning of the first satellite, so badly misunderstood by our political leaders ("silly bauble," "without military significance," "not disturbed an iota," etc.).

Scientific research was not systematically pursued until 1600. Since then it has given man, for the first time in history, a set of verifiable statements of fact found true by all who check them or apply them, anywhere, at any time. This body of knowledge, small at first, has grown and is still growing, like a sum of money constantly reinvested at compound interest.

Scientific research, however, is the search for pure knowledge, and it does not concern itself with applications. The practical application of scientific knowledge in the chemical, mechanical, and finally the electrical industries became more and more important throughout the nineteenth century. Around 1900 there began to appear a new type of activity, which in its turn is now growing at compound interest.

This new type is a deliberate combination of scientific research, industrial design, and marketing. For example, nylon is a new chemical compound; it is also a new yarn, as well as a new style of men's and women's apparel. A transistor embodies a new theory in solid-state physics; it is a new circuit component; and it means hearing aids almost as portable as eyeglasses, among other things. An advance in the understanding of insect respiration permits the development of a new chemical spray which protects crops from insect invasion.

This triple fusion of scientific method, industrial power, and satisfaction of the needs of the common man has proved so successful that many industries, until now completely empirical, are becoming highly scientific, one after another. The most striking example is the age-old art of metallurgy, which had made little progress until the nineteenth century, and which is now being made over by merging it with one of the most sophisticated branches of mathematical physics: quantum mechanics.

This suggests what the power of science means to everyone, but especially to the underdeveloped nations of the world. Progressive leaders in every underdeveloped country know that the techniques for growing food, improving public health, and controlling population growth are now fundamentally understood. They want to change the conditions in their respective countries from misery for most to a decent living for all. They count on science to achieve this. From the widely different examples of contemporary Germany, Japan, and Russia, these leaders see what a school system that teaches science and devotion to the country can do to modernize a nation, even under the severest handicaps. And they ask for competent and friendly foreign experts to help them build four essentials for a modern nation: fertilizer plants, steel mills, teachers' colleges, and medical schools.

A New Culture

Our own problem is to bring the American public to realize what science is. Three main difficulties stand in the way. First, the non-scientist knows little or no science, as a rule, and finds nothing in the history of other cultures to help him conceive its present-day significance. Second, the scientist is usually a specialist in a narrow section of science, and he is much too excited by what is happening there to want to spend time explaining it to others. Finally, science is a difficult study. It is a severe taskmaster: it accepts nothing slipshod. One mistake in the algebra or the chemical analysis, and the answer comes out wrong. This very strictness is one reason why science is so powerful. It is also what makes science understandably unpopular in American high schools.

It is only very recently that our lack of interest in science has become a danger. During the 1930s the United

States could still afford to rely on a very small proportion of gifted volunteers entering science or scientific engineering, because so few industries were scientific and the European output in pure science, on which we relied in good part, was very high. Now we need many more scientists and engineers, because numerous industries make use of science at a very high level and because giant armament industries—a relatively new development—are competing for personnel with production industries of the older type. Also, we must export engineers to underdeveloped countries if we want to spread American influence; quite apart from better reasons, engineers are many times more effective to that end than financial loans. Last, Europe is still groggy from two world wars, and in spite of its high scientific potential and local successes it is no longer spearheading the scientific advances of the world as it once did. Thus we are more than ever dependent on our own strength.

In order to maintain its position of leadership, our nation as a whole must understand the role of science in the present transformation of the world. It must assume responsibility for the changes that are coming. That is not an easy task for a culture which has its roots in Homer and in *Genesis,* and which clung to traditional ideas for twenty-five hundred years before it was rudely shaken by Copernicus, Newton, and Darwin. Each of us will have to bring his contribution to the difficult new problem of living with science, and help find out what ends we wish to serve through this gigantic power that has so recently fallen into our hands.

The word "culture" sometimes is used to mean ways of doing and sometimes to mean ways of thinking. The great cultures of the past have been periods of history when these two meanings fitted together, when a nation looked at what it was doing and saw that it was good. We delight in retrospect in the harmony of thought, art, and life in the Age of Pericles, or of the Sungs, or of Louis XIV. Our urgent need today is for an American culture in which science and technology would play their role as two of the many parts of a harmonious whole.

To this end, a genuine knowledge of the basic sciences must be the possession of every man. The scientist or engineer needs it in order to see his own specialized work in the perspective of an over-all scientific and technical pattern. The non-scientist needs exactly the same basic scientific training because a glance at science from the outside will never give him an understanding of the inner springs of this new intellectual power.

The non-scientists are the men who, out of the findings of the natural and social sciences, must evolve a new culture in the humanistic sense. Not until lawyers, historians, novelists, playwrights, newspapermen, and above all philosophers (who are already leading the way) learn the basic sciences in school and college shall we have the culture this age requires. The more science transforms our ways of life, the greater is the danger that the more literate among us may be unlettered in science. Such men should be thinking ahead of the progress of science, not trailing behind it. To give two examples: Advances that have taken place in the last few years in the widely differing fields of automatic computation and of biochemistry will bring as vast a social revolution as did the steam engine, and they will have a greater philosophical impact than did the theory of evolution. To adapt the general thinking to such new intellectual tools is the duty of "philosophers" in the broadest sense, whereas at present the thrill of seeing such changes emerge is reserved for only a few specialists.

The Task Ahead

There is a huge and urgent task ahead for a new education at all levels, based on the recognition that we live in an unprecedented "age of science." This new education will demand more intellectual work from the educators and from the students, more taxes to pay for school buildings, scholarships, and salaries, more "homework" for all adults in order to keep up with scientific advances—as scientists, engineers, and physicians are already doing.

It almost seems unreasonable to expect a whole nation to impose such strains on itself. Yet it is not unreasonable if we observe that such strains are normal in time of war, and that the Cold War is a war for all that. Our adversary has taken full advantage of the progress of science, not only on the interior front but also in his external propaganda, where he presents himself to the world as the champion of science, the bourgeois countries as sunk in obscurantism. It should be clear to any student of science and of its history that science is an extended episode in the evolution of man, running its course regardless of the opposition of some and the acclaim of others. But in the shorter perspective of our lifetime, a political struggle is going on for the leadership of the world, and it demands the same abdication of the easy-going ways of peace, the same devotion to the over-all interests of the country, that all Americans are willing to give in time of old-fashioned "war." The crisis in our scientific education is only one aspect of this wider problem. A nation is doomed if, in time of war, a majority of its people continue to believe that the basic rule of conduct is one's own immediate interest.

❊ *Margaret Mead*

6

The Gap Between the Scientists and the Others

MARGARET MEAD

Recent attention to the state of science teaching in the United States, and the images of science and of scientists held by American school children and the American public, document a situation which has been developing over the last fifty years—at first slowly and insidiously but since World War II at an accelerated pace. Increasingly, school children and those adults in our society who are not directly concerned with science have come to feel that science is something different and alien, a discipline they neither can nor care to understand. Fifty years ago all educated people had something of a common culture in this realm: almost every college graduate knew at least some science and was capable of appreciating scientific advances. But today there is hardly any sense of community between professional scientists and educated laymen. The scientist, or the schoolboy (it is almost always a boy) who is identified by teachers and fellow students as a future scientist, is set apart by the very nature of his interests and is regarded by *non*-scientists as a person who has restricted himself by choice to the company of other scientists, unsuitable both as marriage partner and as friend.

The troubling aspect of this situation is that matters do not improve appreciably when the image of the scientist is a "positive" rather than a "negative" one. A study made for the American Association for the Advancement of Science showed that a concerted effort to communicate a positive image to school children has had considerable success.* The students could discuss the dedication demanded of a scientist, the years of patient work, the toleration of frustration if after all those years one found one was wrong, the responsibility and integrity demanded of one who was the custodian of truth, of mankind's welfare, and of the safety of his country, and who must be immune to temptation and relentlessly devoted to his work.

But this image of dedication does not attract students; like the images drawn from the mass media of the scientist as a mad, godless "brain" or a sort of "sorcerer's apprentice," it repels. In part this repulsion may be attributed to the shift of young Americans' values away from areas of work that require long-time commitments and toward their present preference for home life over careers. But also, in

* See Margaret Mead and Rhoda Metraux, "Image of the Scientist among High-School Students," *Science,* August 30, 1957, pp. 384-390.

an age when young people are seeking for warmer, more alive human relations, account must be taken of the extent to which scientific work was pictured by the high-school students examined in our study as "uninteresting," "boring," and concerned mainly with dead things. For such categories as "organic" and "inorganic," today's students have substituted "alive" and "dead," and they see science as concerned primarily with the latter. In addition to mathematics, physics, and chemistry, the so-called life sciences are seen as concerned with dead animals (or parts of dead animals), or with living animals that are caged ready for death, or with organ systems of the human body abstracted for study. Scientists in other branches are seen as concerned with stars outside the living world, or, most significantly, with archaeology and the digging up of masses of potsherds in dead cities. The absence from this picture of the study of living things outside a laboratory and of any knowledge of or interest in the human sciences is striking. These images, held by students who are the coming citizenry of a country which must make ever greater efforts to find and to educate scientists and scientific workers of all kinds, to finance scientific research, and to legislate in a world transformed by science, present a serious challenge. Although a few steps to alter these images can be taken through the mass media, the causes lie deeper and call for more intensive remedies.

We are, in fact, in danger of developing—as other civilizations before us have developed—special esoteric groups who can communicate only with each other and who can accept as neophytes and apprentices only those individuals whose intellectual abilities, temperamental bents, and motivations are like their own. A schismogenic process is under way that is self-perpetuating and self-aggravating.

When the American high-school system shifted from

the school that prepared a small number of students for the university and that had a curriculum in which mathematics and science were taken for granted to the school that had the diversified curriculum believed necessary for the education of all young people, the way was open for the operation of a selective process within which only those students who "liked" or were "good at" mathematics and physics elected these subjects. Classes no longer consist of a group of intelligent students with a wide scatter of abilities whose mental processes include a variety of sensory modalities and whose preferences range from the classics to the construction of machine models. Rather, in each class the students are self-selected, vocationally guided, parent-pushed, and university-pulled in their choices. They choose against science unless they have the specific type of mind believed to be "good at" it and the kind of single-minded devotion to the smell of a laboratory that leads them to prefer working there, not only to going to the ball field or the corner drugstore but also to working in a library or studio.

By World War II the future young scientist was well separated from his fellows, and at the college level the engineering student was expected to have difficulty in writing English and to abhor the humanities, just as students of the humanities were expected not to understand and to abhor mathematics and science. By this time, too, the steadily widening distance between students in the different fields was beginning to show up in the elementary schools. And in colleges and universities there was a hardening of the reaction formations so apparent in the humanities today, in which students of the humanities, as they lost their hold on contemporary developments in science, began to stress their monopoly of eternal values.

The situation in which small groups of specialists

elaborate esoteric knowledge which becomes progressively more inaccessible to the rest of society is not historically unique. It has existed in many societies as a concomitant of the invention of script, especially of a script for whose use a detailed knowledge of a long classical tradition was necessary—as in Chinese or Arabic. Then the selection of a few children in each generation with the necessary special abilities assured not only the continuance of the tradition but also its ingrown feeding upon itself, until conquest, revolution, or violent culture contact toppled the system and shook the esoteric priests or teachers out of their isolation.

But there are two aspects of twentieth-century science for which we have no historical parallels. The first is its progressive growth, each step of which makes other steps possible—in contradistinction to learning in past civilizations, which did not involve the built-in possibility of such continuous cumulative discovery. The second—and even more important—aspect is the changed pace of discovery. Both of these aspects, peculiar to modern science, have been conducive to the development of a style of cultural transmission that departs sharply from the age-old methods man has used to consolidate the gains and insights of one generation by teaching them—sometimes articulately and sometimes inarticulately—to the next generation.

It has been characteristic of all earlier forms of cultural transmission that new intellectual acquisitions—such as script, mathematical calculation, prosody—have been taught in face-to-face situations by adults who knew to children and adolescents who did not yet know. However great was the delight of communicating a new advance to one's highly trained peers, the peers were few enough, the need for apprentice learners was great enough, and the pace of discovery was slow enough so that the process of teaching was an integral part of the development of theory and practice. However exciting and rewarding a shorthand conversation with the few adults who could immediately grasp a new discovery through the use of specialized vocabularies and precise language systems, however strong the tendency toward isolation and the proliferation of coding, these were compensated by the need and time to teach, generation by generation.

In the last twenty years this situation has changed abruptly. As the several sciences have begun to grow at an unprecedented rate, the distance from one major advance to the next is often reduced from fifty to five years. And now, instead of teaching a widely selected, intelligent student audience, more and more the young scientists are communicating to each other, horizontally, in highly specialized languages, material so new that publication is not rapid enough to encompass it. We now have the working conference, in which face-to-face communication in all sensory modalities takes place among a group of young scientists, all of whom speak the same specialized and specific language, which is supplemented by the type of face-to-face learning that once existed between master and apprentice but that today tends to get lost in formal school or university situations. Tremendous scientific progress is possible in this type of communication, where all the participants are adult, highly educated, and specialized, and where the "teacher" often is younger than the "pupil"—where, in fact, such distinctions may become quite meaningless in the mutual excitement of scientific advance. (But, meanwhile, even the well-trained science teacher, cut off from these rapid interchanges in the laboratory, becomes steadily more isolated and his teaching more static in quality.)

All of us who cherish the change in pace made possible by this new kind of horizontal, face-to-face, multimodal transmission, which works even across national boundaries, inevitably will guard jealously any attempt that would seem to slow up this intoxicating process. But now we must find new educational and communication devices which will not sacrifice this new high level of specialized communication and yet will protect our society and all the intellectual disciplines within it from the schismatic effects of too great a separation of thought patterns, language, and interest between the specialized practitioners of a scientific or humane discipline and those who are laymen in the particular field.

At present there is a growing effort to identify those children who display a particular type of intellectual ability and to provide a special education for them. This is as necessary in the clinically oriented human sciences as in the physical sciences. But this effort ought to be complemented by an effort to teach as fully as possible the advances in any field to other young children, and to adolescents and educated adults as well, so that we may have scientifically literate citizens and parents.

There is yet another urgency in spreading the understanding of these advances. The process of vertical communication of results arrived at by horizontal, face-to-face adult learning will alter the vocabulary and syntax of the *communicators*. Thus they will be the more able to transmit what they know, and they themselves will keep in closer touch with the other specialties of our highly specialized societies. Any language taught only by adults to adults—or to children as if they were adults—becomes in certain respects "dead." It fails to enlist recruits, it may lose its productivity, and it serves in the end primarily to separate those who know it from those who do not. In contrast, any language that is taught to all children attains a multimodal comprehensiveness that makes it a suitable vehicle for the thought of not only the highly intelligent but also the moderately intelligent and the deviantly endowed person. By insisting that all children should be taught recent advances in a particular discipline, we can set up an automatic corrective system for the dangerous intellectual divergencies of vocabulary and knowledge within our society.

In this way we shall add to the disjunctive, horizontal, face-to-face adult learning the steps peculiar to the developing human mind, and so we shall steadily make each advance available to the whole society. The existence of the mass media makes it possible to combine this purposeful inclusion of primary learning by children with a different type of learning for adults. For in broadcasts and telecasts into the home, where the learning child and the learning and immediately teaching adult are included in the same audience, the lay adult is given a condensed experience of primary learning—of how he would have learned some unfamiliar subject if he had learned it as a child. (This is comparable to the way adult immigrants acquire a better understanding of the culture of the new country as they participate in their children's learning. The isolated, childless immigrant may quickly acquire an apparent mastery of the new culture, but his knowledge will lack depth and be difficult to transmit—just as the specialized scientific languages, which are learned only by specialized intelligences, become steadily less communicable and less learnable, and so more and more isolate those who speak them from their fellow men who do not.)

Any subject, no matter how abstract, how inanimate, how remote from the ordinary affairs of men, remains lively and growing if taught to young children who are themselves growing by leaps and bounds, hungering and thirsting after knowledge of the world around them. To children, an understanding of the world around them is as essential as is the tender loving care that, during this century, has been so exclusively emphasized in discussions of early childhood education. Whereas the apathy of older students whose minds have been dulled by years of weary, repetitive teaching may repel the scientist who lives on the frontiers of new discovery, the questions of young children can challenge the most erudite to prodigies of exposition of hitherto recalcitrant materials. The language of science will then become—for everyday use—a natural language, redundant, wide in scope, deeply rooted in many kinds of human experience and many levels of human abilities.

❋ *Warren Weaver*

❋ *Fletcher G. Watson*

7
Science Teaching

WARREN WEAVER

The first purpose of science education as a part of total education is to furnish a chance to young people to get some insight concerning the physical and living world about them: the beauty, the discoverability, and the dependability of its order; the elegance of its logical structure; and the fun of discovering and sensing all of this. Such understanding involves intellectual satisfaction. There is also aesthetic satisfaction in appreciating the scientifically revealed unity behind the variety of nature. Coleridge's conception—that the essence of beauty was the discovery of unity within and behind obvious variety—has always been useful and appealing, at least to me, because, as Bronowski has so effectively pointed out, it unifies the positions of the creative artist and the creative scientist.

Secondly, it seems to me that a main purpose of science teaching is to give young people a chance to get training in and experience with a method of thought, an attitude of mind, a technique of inquiry which has had unparalleled success in certain areas of human experience, and which certainly has wide usefulness in many other areas of human experience to which it has as yet not been fully applied. This technique of inquiry—sometimes referred to, a little overpompously, as "scientific method"—seems to me best captured in the very simple phrase "disciplined and competently served curiosity." It is not just curiosity; it is *disciplined* curiosity, and it is *competently served* curiosity. This to me is the essence of the manner in which the scientific mind deals with the world.

The third of the major purposes—putting these in a descending order—is to give young people a chance to learn a *part* of the modern world of idea and action without which no one can hope to be an intelligent citizen in a modern, free society. Science has become so inextricably interwoven into the fabric of our culture, so inescapably necessary to the maintenance of our level of economic activity, so essentially involved in the protection of our freedom, that no citizen of our democratically determined society can participate effectively in decisions unless he has some reasonable knowledge concerning science.

Fourthly, another general purpose of science education is to give these young people a chance to become interested in science as a profession. At the early stages of education it should be an opportunity and nothing beyond an opportunity. I think that all of the pressures, all of the pleas—perhaps I should even use the phrase "all of the bribes"—should come at a later stage. I do not think that it is healthy to reach down into the high-school level and try to persuade youngsters at that stage to become scientists by offering them scholarships that carry them all the way through.

I do not have very much confidence in the accuracy or the reliability of any of the present attempts to make early assessments of aptitudes and capacities. I doubt that one can do this safely at the ages of thirteen or fourteen, and I consider it also unfair to the youngsters. They ought to have access to or knowledge of a great many other interesting and rewarding and wonderful ways in which they can spend their lives. A freshman in high school cannot be expected to know what it means to be a musicologist, and perhaps his career should be that of a musicologist, or a philosopher, or a student of linguistics, or a minister. It is important to preserve such variety in our culture as will give every pursuit its own rich chance to recruit students. I do not like to see the scientists, and particularly, the armed services, go down to low levels of education and shake the money bag in front of bright youngsters with an invitation to become scientists.

In addition to the purposes of science education considered so far, there is, of course, the exceedingly important purpose of making competent scientists out of those individuals who have become dedicated professional candidates. When does the decision of dedication occur? I don't know. An estimate based on my own years of teaching experience places this decision somewhere about the sophomore or junior year of university, a good deal earlier with a few scientists, and occasionally later. When they do take scientific vows, when they become professionally dedicated, then it is not too difficult to make scientists out of these students. I think we do this job well, on the whole. It is the rest, the broader job, that I do not think we do very well.

How can the four main purposes that I have mentioned be served? In a presentation that is intended to stimulate thinking and discussion, I shall not hesitate to offer a very scattered set of comments.

First, I believe that science teaching in the elementary and secondary schools, and science teaching to non-scientists in the early years of college and university, is generally rather poorly done at present. One important reason is that there are far too few good teachers—and I should like to emphasize most strongly that this is not their fault. It is our fault. We must restore scholarship to elementary teaching. We must restore prestige to elementary teaching. And we must start paying teachers a decent amount of money.

The most important of these interlocking needs, in my judgment, is improvement on the scholarly level. Some of us feel rather strongly that the less enlightened leaders among professional educators have placed an unfortunate emphasis on method, as apart from, and sometimes even contrasted with, content. We feel that this relative emphasis should be reversed. And while we wish children to develop all their potentialities, we feel that a disciplined development of their minds is a central and essential core of the process.

Secondly, the curriculum material has not been very good. Much of it is antiquated. Most of the texts try to cover altogether too much factual material, and many make such a frantic and superficial effort to interest that they succeed only in being dull. I do not think that these texts are too hard; on the contrary, they are not hard enough. They underestimate the ability of the young students and do not treat them with enough respect. Students are capable of responding to much harder material if it is stimulating, solidly interesting, and intellectually respectable.

Some of the mathematics taught in high school, for example, is, I think, rather dreadful; I do not blame students for being bored with it. Surely, we should teach more aspects of mathematics that are modern, and interesting, and good. Much as we emphasize the beautiful, logical structure of classical geometry—the austere and elegant procession of pure assumption, pure logic, and the unexpected unfolding of rich result—how many times in a lifetime has anyone confronted a situation in which he had to draw a really strict logical conclusion? I, for one, do not use syllogisms around my house very much. As a matter of fact, the characteristic situation in life is one in which we must make decisions without having enough knowledge of the facts to permit the application of simple logic. Mathematics has a great deal to say about such real situations, but in a field that most young students do not encounter, namely in probability theory and statistics.

There are, happily, some promising new movements on the improvement of the curriculum. I think one of the best attempts is the so-called Physical Science Study Committee, originally at the Massachusetts Institute of Technology, under whose auspices and leadership a group of first-rate working physicists is trying to produce a curriculum, a textbook, new experimental procedures, and a whole battery of teaching aids for high-school physics courses. There are other interesting developments of this sort. The mathematicians, too, have several such projects. The biologists and the chemists are still lagging a bit behind, but there is hope on the horizon. And the best part of the hope is that good working scientists are now taking part in this job.

I fully realize that I am an ancient and dyspeptic character who has not been a teacher for twenty-six years and who is therefore out of date. I realize that I may not be properly informed about the present happy facts concerning schools of education and the curricula imposed on those who must have a teacher's certificate. But, suspecting that there is still a point to the protest, I feel compelled to state, as emphatically as I can, my conviction that the primary intellectual qualification of a teacher is to know the subject being taught. I would put second in importance that the teacher have a broad general education. A useful and important part of general education for a prospective teacher of course would be some knowledge of the history of education. To keep this subject in proper balance, it should be taught by scholars in a department of history. Only in third place would I agree that there is doubtless a modest volume of dependable and useful information about methods of teaching; when directed to a large body of average individuals, a little instruction concerning methods should be moderately useful.

It is one of our noblest attributes as a people that we are determined that everyone in this country be born to an equal opportunity. But in far too many cases we have assumed that equal opportunity means uniformity of treatment. I believe we should have far greater variety in what we offer as educational opportunity for our youths. Some are capable of going three times as fast as our ordinary curriculum goes; we should allow this and make it possible. There are many others who should receive an education that recognizes the value and the dignity of the good artisan. The need, in short, is for both excellence and variety in our educational pattern—not least, in our science curricula.

FLETCHER G. WATSON

For any long-term influence upon the attitudes of future adults, programs of instruction must be consistent from the first day of school in the kindergarten to the day of high-school graduation. One cannot expect to produce educated men and women of the twentieth century by exposure to a year or two out of a layer cake of specialized science courses in the last few years of the high school. These courses are elective, generally fairly difficult, often mathematical and abstract. The result is that, even among the children who stay in school long enough to reach these grades, many choose other courses of greater appeal or those in which their chances of academic survival are greater. While drastic changes in these final courses are desirable, a receptive attitude toward science in the majority of pupils must be developed through courses offered much earlier.

The literally "wonderful" aspects of the world around us must become known to the child. He must also be encouraged to "wonder" about it, and then to discover the amazing human attribute of being able to pattern events into predictive systems. This takes time and sympathetic assistance from emotionally secure and informed teachers. A truly staggering task, then, is the introduction of effective science programs in our elementary schools, which employ some 800,000 teachers, many of whom feel uninformed in science and frightened by the very term.

Equally important in the development of informed and sympathetic children is their instruction in the junior-high-school years, grades 7, 8, and 9. For these more mature children, different approaches and more specific activities and inquiries are essential. Yet materials of instruction in science are often woefully lacking, even when adequate time for science instruction is provided in these grades. As these are the years in which children begin to become realistic about possible careers and when they must choose among alternate programs in the high school, effective and stimulating teaching is surely desirable. Unfortunately, such teaching is not commonly offered.

For at least ten years we have annually been short several thousand able new teachers in science and in mathematics. Our accumulated deficit throughout the nation is now so enormous that even temporary stopgap measures are no longer adequate. This tragic condition occurs just when the importance of science and mathematics is being stressed and a sharp increase in enrollments is occurring in the high schools.

While time for planning and reflection is essential for creative teaching, our teachers are harried with too many classes and much busy-work (milk money, selling tickets, supervising study halls or lunchrooms, etc.). In addition, many are obliged to carry a second job to keep their families solvent. Our platitudinous talk about the importance of education and schooling does not square with the actualities of the grudging and inadequate support we provide for education.

We shall probably not be able to catch up with the total need for teachers for some decades, but much can be done to make more effective use of those we have and to improve the others. Obviously, as the 1958 Parliament of Science concluded, "we must compensate teachers at levels which reflect the degree to which the destiny of the nation depends on teaching of the highest quality." This would mean that salaries must be raised to the point that schools can expect and demand the full-time, year-round interests and activities of all teachers. No longer is teaching as a part-time sinecure adequate for the needs of the country.

Able teachers must be used where they are needed most. Too often teachers have their daily schedule filled out with classes in subjects in which they have neither knowledge nor interest. Once they sign a contract to "teach at the assignment of the school department," they have no legal recourse against misusage. Certification laws, vague and inadequate as they are as protection for the children, become a mockery when unpoliced and uninforced. Among needed positive action is the employment of able teachers who are available only for part-time instruction.

Competent teachers also might be shared among several schools rather than assigned to chore work to fill out a day's schedule. Furthermore, much of the routine work (grading papers, policing lunchrooms and study halls, etc.) can be assigned to less highly trained employees.

Finally, teacher improvement, now usually left to the individual on his own time and expense, except for such limited programs as those financed by the National Science Foundation, must become of intellectual and financial concern to the schools. While American industry expends over five billion dollars annually for the further education of its employees, only rarely does a school system have a mechanism for financing further study by its teachers. Where inevitably teachers differ in their special skills and abilities, the more able ones must be used to help those less able—and they must be properly compensated for this activity.

❋ *David Riesman*

8

The College Professor

DAVID RIESMAN

In this informal consideration of the academic career in America today, I shall describe the pattern of recruitment into the academic profession, and suggest some of the ways in which this pattern has changed in the last generation or so. I shall discuss attitudes toward the college professor, and the sources of his morale. Finally, I shall become more personal in considering some of the perils and opportunities of teaching in the social sciences.

We now have an extraordinary literature about the academic man written by novelists. Whereas novels about college life once tended to concentrate on the students (as in George Weller's novel of Harvard, or Percy Marks' of Brown), today the faculty, and especially the administration, become fair game. The colleges apt to be pilloried in such accounts are those that harbor writers, and the image of the academic profession presented in these novels—I think of the work of Mary McCarthy, of Randall Jarrell, of Stringfellow Barr—is taken from a segment of the loftier, the more experimental, and perhaps the more bizarre institutions, not, as one might conclude from innocent reading, the more benighted ones.

Sociologists have seldom turned their attention to our institutions of higher learning or focused on the professor as a subject of study. But they are now beginning to study colleges along with other major social institutions. There will shortly appear a study (commissioned by the Fund for the Republic) based on interviews with 2500 social science professors in a sample of accredited colleges and universities. Substantial investigations of colleges as institutions are under way at Yale, Vassar, and the University of California at Berkeley.

No one should underestimate the difficulty of research in an academic community. The "natives" are often by definition individualists who will not necessarily stand still to be interviewed! Colleges themselves, of course, do not stand still. They have a rapid turnover of students and frequently of faculty and administration; their purposes, their patterns of recruitment, and the destinations of their students alter even while their professed ideology may remain unchanged. Moreover, colleges, unlike big industry or big government, have not got money to spare for research (although one or two universities are developing self-study units); in contrast with many industries, colleges are doing their long-range planning mainly on a play-it-by-ear basis. On many campuses the social science departments are not highly regarded, and they would be even more deprecated if they should start to study what goes on in the classrooms of colleagues or outside of classes.

For all these reasons, therefore, not very much is known about what happens in college (or, for that matter, prior to college) that might lead a young person to consider an academic career or to reject it. Two studies recently done by faculty members at Wesleyan University, however, give a rough and approximate sense as to which colleges and which types of colleges send people into graduate schools of arts and sciences and into more or less productive careers as scholars. These studies have surprised many educators, for they show that many of the leading and most distinguished universities (Harvard, Yale, and Princeton, for example) have turned out relatively few scientists in proportion to their graduates, whereas a number of small and often impoverished liberal arts colleges, primarily in the Middle West, have turned out, in proportion to their enrollments, a great many. Some of the colleges with distinguished records in this respect are well known for their educational venturesomeness—for example Antioch, Swarthmore, Reed, and the University of Chicago. But others in this group, such as Grinnell, Hope, Linfield, Kalamazoo, and Wabash, are either not widely known or not in any way experimental; many still retain their ties with the Protestant churches under whose auspices they were founded. The Catholic colleges have a very poor record, according to the criteria of the Wesleyan researches, as have the Southern universities save in one or two fields.

There are problems in interpreting these data, especially concerning the records of the state universities, whose undergraduate bodies are not strictly comparable to those of most liberal arts colleges. Nevertheless, a few tentative conclusions can be drawn.

In the first place, while the Bible Belt of the South has produced mathematicians and poets, it is from the Protestant but not Fundamentalist colleges of the North that a relatively large proportion of natural and social scientists have come. The affinity between some versions of Protestantism and scientific work has often been remarked. For the less hierarchical churches of Protestantism ask the individual to determine his own relation to God—to search out his own cosmology. As it becomes secularized, this search may take scholarly form, or indeed, as often happened in the sixteenth and seventeenth centuries, one may try to reconcile science and religion through scientific work or even to show the glory of God through the shape of the heavens. Correspondingly, as is well known, a number of men in the older generation of social scientists are ministers' sons who have found in a scientific career a way of sublimating or transcending some of the conflicts between the scientific world view and the Protestant one with which they grew up.

In the second place, the colleges producing notable scholars often have drawn on rural and small-town constituencies and on the lower middle class in general. That is, they have drawn on people who, when they landed in college, did not have a very differentiated or complex idea of what the world offered. If they came from a farm, for example, they might have been able to see themselves getting away from a plow only by becoming tractor salesmen or county agents—and then they discovered in college that they could become botanists or biologists or teachers in the agricultural colleges. This opportunity proved to be a way of getting off the farm while still retaining a tie with rural life and its values. Similarly, a bright boy from an impoverished background might land in college without ever having heard that one can make a living as a physiologist or an astronomer. But he might have had the luck to encounter in college a teacher who was doing just that. If this teacher befriended him and offered him, so to speak, the key to the laboratory, the student might end up in his master's shoes—a captive of the field whose horizons were thus opened to him.

In other words, the very lack of cosmopolitanism of some of these colleges (especially perhaps in the Middle West), and the lack of cosmopolitanism of the students who in the past went there, meant that a teacher of even moderate quality and interest in his students could accumulate disciples quite readily. Conversely, the inferior record of the great cosmopolitan universities in recruiting undergraduates into academic careers has in my judgment been partly due to the fact that students who went there have had many other choices in mind. (Likewise their social science professors, busy with graduate students, with consultantships, and with all the opportunities and temptations of a metropolis, have also had other alternatives to looking for disciples among their undergraduate students.) That is, such students have found other ways to spend their time, even other intellectual ways, than in the laboratory or in the office of the favorite professor. They could envisage themselves (assuming they eschewed business) becoming diplomats, journalists, or TV script-writers, along with a thousand other opportunities offered by the big city. Moreover, students coming from homes where the parents themselves attended college on the one hand might think of an academic career as a downward step in social mobility (as against the upward step that it was for the farmer's son or the person from the lower or the lower middle class), and on the other hand, having already been somewhat exposed to academic values, they might not be captured as readily by them through the medium of an exciting professor.

We can approach some of these same matters in another way by reminding ourselves of the experience many prospective scholars have had in their high-school years. Except at a rare urban institution like Bronx Science or Hunter College High School in New York City, a boy or girl who is headed for science and scholarship feels out of place in high school. He is bookish when nobody is bookish, or he putters with chemicals when other boys are out on a date or at sports. He feels alienated and alone. He ends up as valedictorian, recognized only in that ambiguous way. Then he may be picked up by a recruiter from a college that is looking for a few scholars to balance its athletes and Good Time Charlies. At last he may find himself not alone; for the first time in his life he has allies and, in the person of a teacher or two, even models and sponsors. Moreover, the teachers can become allies against parents who have often felt troubled about their child's unworldliness or bookishness or "queerness." Henceforth the student may be set for life in a new mold and a new career, possibly discovering that the things that were regarded as vices in his high-school years now turn out to be virtues.

The Wesleyan studies, and consequently the interpretation I have drawn of them here, are necessarily already out of date, since the scholars whose careers were sufficiently advanced for statistical follow-up mainly attended college in the period between the two world wars. As readers of Sinclair Lewis and Sherwood Anderson will remember, this was a period in which, in the smaller and especially in the Midwestern communities from which scholars came, there were often strong anti-intellectual currents; in which the dominant values were those of money, power, swank, Republican respectability, and practicality. In such a climate of opinion, it was understandable that professors should have been regarded as stuffy, as not quite manly, as occupants of an ivory tower that probably needed dusting. (Or alternatively, professors were regarded as radical cranks who could be trusted neither to teach the right brand of economics or theology nor to provide models for the success-prone student.) In that climate of opinion, the handful of alienated students would naturally find themselves sympathizing with their college professors; and conversely, the professors would themselves be looking for recruits among the students as hostages against the culture of Babbittry around them.

In such a pattern of recruitment into academia, it was plain that not many would be "called"—that is, find so peripheral a calling to their liking. As a result, professors could and did spend their time with a few students and tried to deal with the rest by liberally distributing gentlemanly C's. And so it was that those boys from the lower or humbler strata who aspired to become professors would be slowly groomed for that recondite elite.

The Spread of "Collegiate" Values

The present situation is quite different. College faculties have been expanding too fast to permit the slow and careful grooming of a few hand-picked scholars. Indeed, in some fields men now may become professors at an age at which they would once still have been teaching assistants. A bright assistant professor today may be more sought out than a famous full professor even ten or fifteen years ago. The competition of the three state universities of Michigan is only an example of the raiding that goes on quite generally. A great many factors have cooperated to produce these changes. The country is richer, which means that it needs fewer people to tend its farms and factories and can locate more of them in the professions and in the other more or less intellectual careers such as communications, management, and teaching itself. It can also afford to keep people in school longer and to send many more of them to college. And people have more leisure in which to absorb some of the "cultural" values previously associated with attending college; the better large-circulation magazines and networks, paper books, and "art" movies spread intellectual values, at least sporadically. More people travel, visit Europe, read *Gourmet* magazine, and in general feel entitled to follow the style set by those who have attended college. In the process America has become not only more urbanized but also in many ways more urbane.

Correspondingly, I believe that the small liberal-arts colleges of the Midwest may not show up as the great recruiting grounds for scholars twenty years from now. The sharpness of the conflict between science and religion is attenuated today; it seldom drives people into a creative tension that results in their becoming biologists or psychologists. Nor do the remaining Protestant colleges seem to me very ascetic. Moreover, there are fewer boys who want to get off the farm by becoming teachers; there are fewer farms, and those that remain are often large and complex businesses.

At the same time, I am inclined to think that the big cosmopolitan universities are no longer channeling their best students largely into such socially approved careers as law, medicine, business, and the diplomatic service, but are increasingly serving also to recruit scholars. This is in part because their student bodies have become less "social," and include many relatively poor boys who once would have gone to local and often to denominational colleges. For example, the Minneapolis *Star & Tribune* now sends small-town newsboys on scholarships to Yale and Harvard, boys who in an earlier day, if they had attended college at all, would have gone to Carleton or St. Olaf's or Gustavus Adolphus or the University of Minnesota. One result may be that, in the future, institutions such as Carleton and St. Olaf's will not show up quite so well as they did in the Wesleyan studies, while Harvard and Yale may show up somewhat better.

For the small-town newsboy skimmed out of the Midwest by the increasing talent hunts of the national universities, the transition from high school to college may still be very sharp and dramatic, even traumatic. But as I have implied, college is not so sharp a break from home as it once was for the many youngsters whose high schools have already anticipated college (as some of the wealthy suburban high schools do) and whose parents are themselves collegiate. It follows that, while youngsters who are *entirely* given over to intellectual or artistic values may still feel quite alone, those who have a partial but not exclusive interest in those values no longer need feel, either at home or in high school, that they belong to a minority culture. While, as we all know, a great deal has been said about anti-intellectualism in America during the McCarthy years and later, it is at least arguable that these very attacks on the intellectuals are, *inter alia,* a response to their rising power in a society which more and more requires intellectuals, or at least educated specialists, to get its work done. Whereas once bankers were hated and feared in part because they controlled the "mystery" of the gold standard, today scientists and intellectuals control the relevant mysteries, and, as Edward Shils points out in *The Torment of Secrecy,* give rise to analogous fears of domination and analogous opportunities for demagogic attack.

The increasing attention to intellectual values is especially striking among some businessmen. In recent years many large corporations have instituted executive training programs so that their middle-management people, who have often been trained as engineers or as business administration majors, can transcend narrow professional horizons and model themselves on the going version of the industrial statesman. And industrial statesmanship today takes a great interest in colleges, because increasingly businessmen do go to college, use colleges for purposes of executive training and development, and meet professors as consultants, market researchers, and social equals. The hope Justice Brandeis expressed before the First World War—that business would become a profession—approaches reality as managers become increasingly aware of the need to handle complex data in making decisions rather than playing hunches or following tradition. Despite ritualistic speeches on ceremonial occasions, big business has become much less Philistine and much less hostile to intellectual values. (Little business, including farming, is an entirely different story.)

These developments render somewhat paradoxical the attitudes I find among many of my own students and those at other leading colleges: namely, a posture of contempt for business and a belief that, in contrast, teaching offers respectability and even integrity. (I should make clear that this is not a political contempt, for these students are very rarely directly political; it is rather a cultural, moral, or intellectual contempt.) Some students come to this outlook because they harbor aristocratic values and look down on businessmen as the English gentry traditionally looked down on people in trade. But others are themselves the children of small businessmen, and they have overgeneralized from their parental occupations, with their often rapacious ethics and lack of intellectual range, to the large managerial businesses I have been describing. That is, small business is still competitive, still what Veblen would have called a "pecuniary" occupation, in the sense that it deals with bargains and mere tricks of trade, not with large engineering and industrial conceptions. The result is that the younger generation, seeking not only social but also intellectual mobility away from the parental small business (and, in this group, increasingly coming from Jewish and Catholic families as well as Protestant ones), have their eye on the professions as the road to status and opportunity. Academic careers then become alternatives either as belated second choices—for instance, a student ending up as a biologist because he could not get into medical school—or as first choices, decided with the blessing of parents and peers.

Many of the parents of these young people, whose values contrast with those of an older day, do not want them in the business but rather out of the business—although, of course, still self-supporting. And many of the occupations that once would have raised the parents' protest today may seem a golden opportunity. Van Cliburn may persuade them that the piano is not a road to ruin; Charles Van Doren may persuade them that one can make money as an English teacher. Less and less are professors regarded as members of a small, deviant but semi-elite group—although, as we shall see, those who teach the humanities often so consider themselves—but rather as people who have gone into a business that isn't business.

Certainly, as I have said, the life of the businessman and the life of the professor become less and less distinct. The professor is no longer to be regarded as a stuffy fellow.

He has become a man of the world, perhaps traveling on an expense account, attending a conference in Washington one day and flying to a UNESCO meeting in Paris the next. In honor of his new status, novels now portray him as having sex appeal and even a lurid sex life. As universities become bigger and bigger, it is hard for them not to judge their output by business standards or at least bureaucratic ones (as I am told the Michigan legislature has recently been judging the educational institutions of that state). I have heard professors in the social sciences pass judgment on each other in terms that would not be different if they were engaged in production control. They speak of a man's "output" or his "productivity" as measurable and even quantifiable things, and yards of print take the place of foot-pounds or B.T.U. (Natural scientists, having the most money, are also exposed to these tendencies.)

Many professors in classics, in literature, in history, and in the humanities generally believe themselves to espouse the traditional academic values. But in the process of homogenization of values that I have been describing, this becomes more and more difficult, and we see the paradox that some of the embattled humanists engage in "selling" the humanities as good for whatever ails the U.S.A. with the same public-relations fervor that they deprecate in Madison Avenue. The social sciences stand somewhere between, trying to drag an uneasy foot out of the humanities while not quite managing to locate the other foot in the profitable camp of the natural sciences. But in fact, as I shall try to indicate, it is impossible to tell whether a man is a humanist or not by the label of his discipline; and I have seen a number of colleges where the anthropologists are more humanistic than the teachers of English, the physicists by far more humanistic than the economists or the sociologists.

As to these comparisons, however, I should add a word of caution against pushing them too far. One of the characteristics of any older generation, markedly of the present older generation in the universities, is to talk about the good old days. In general, it is safe to say that the old days were not so good and not so different as most of us believe. There were plenty of professors at the turn of the century who espoused business values, just as there were many ministers who did, and they did so with a vulgarity that one could hardly match today among businessmen. Correspondingly, there were other professors a generation ago who really were stuffy, pretentious without being literate, erudite without understanding, pedantic without being critical. Along with the general rise in standards of education, I believe college teachers have become brighter and better.

The Distaff Side

So far, I have emphasized patterns of recruitment of college teachers in terms of the backgrounds from which they come and the collegiate atmospheres in which they are nurtured. Before turning to some of the reasons why I have chosen to be a teacher myself (in spite of all the misgivings implicit in what I have said), I want to say something about the problems posed for the college professor as a family man. Professors get many of the benefits of the academic culture while their wives are saddled with many of the drudgeries. In my observation, it is less the professors who suffer from low incomes than their wives and children—especially their wives; and, of course, the wives and children make the professor suffer. This is so in spite of the fact that many men who are teaching today and complaining about their salaries are doing ever so much better than their parents ever dreamed of doing. For the trouble is that their own rise, while considerable in absolute terms, is not great relative to the rise of others who have been equally mobile but have gone into different fields. Moreover, while the gap that once separated an instructor's salary from that of a full professor has been greatly reduced, there is still substantial difference in affluence within a university community, where the professors at the law school and at the medical school, and probably at the business school, may be getting as much as twice the salary of those in medieval history, while the professors of economics and sociology may be more than doubling their salaries with consulting fees—the academic form of moonlighting.

Such moonlighting, furthermore, represents another peril of the academic life: namely, that teaching less and less provides the traditional advantage of long vacations and much time at home with wife and children. Wealthy lawyers, industrialists, or doctors, who work long hours and see little of their families, at least can feel that, after all, they bring home excellent salaries and fees, as well as high community status, even if they bring themselves home too little and too late. The professor, however, belongs to that stratum of society that has come to believe that the father should play a role in the home and in bringing up his family: it is one of his do-it-yourself activities. He is protected neither by his high salary nor by his traditional values against feeling himself inadequate as a parent when he can neither educate his children as he would like nor see as much of them as they might like. For increasingly his work is never done. His reading and research get crowded into the corners of the day. His position is vulnerable to all sorts of demands, and to protect himself he has to be more calculating or more incompetent than he knows how to be or cares to be or can afford to be. In addition, his wife shares his worries, and he must share hers. Hence, while in principle the professor still has more time than most professional men to spend at home, including the long summer vacation, much of this time in fact is spent either earning money to pay the plumber or working like a plumber. Indeed, the college professor must often feel that he is training his students for the "new leisure" while not enjoying much of it himself.

The Search for Colleagues

What is perhaps most characteristic in the work of the college professor is not his relatively low salary (that of schoolteachers, social workers, and ministers is often considerably lower) nor the fact that he grows older while his charges do not (for he shares this characteristic also with schoolteachers, pediatricians, policemen, and wardens), but rather that he sets his own goals; the goals are not given by an institution. This is certainly so in the penumbra of freedom beyond his regular teaching and other curricular duties, and even these duties, the higher one rises in the system, are defined in terms of one's own aims and definitions of the situation. It is this that makes the professor kin to the artist or writer, who is often seeking to create his own institutional norms; and it is this freedom, I am sure, that attracts many to the profession. It certainly attracts me.

But no one should underestimate the miseries of having to set one's own goals. It is much easier, in my observation, to meet a payroll or to try a case than to do research or to try to help students learn something. For creative intellectual work is never done, and it is certainly never done to one's own satisfaction. To be sure, professors and creative people, like other people, try to find outside judgments in order to avoid having to set their own goals. Sometimes these judgments are found in the tradition of work within the confines of a particular school. By adding a little bit to the enormous, expanding edge of what is known, one can feel that one is part of a cumulative process that is primarily a collective enterprise; and in such work, one will be supported by the contemporaries of one's own school and tradition of work (whether this rests on a particular method or on a particular substantive area), and one will seldom have to ask oneself whether that tradition is itself worth while—especially since, guided by that tradition, one can keep turning out papers and graduate students. I should add, of course, that much useful work is done in this way by conscientious craftsmen, and some careful work by opportunists—in fact, the motives of most of us, for most of the time, are mixed and have very little to do with the truth of our discoveries.

The scholar who rejects this course may seek or find an audience in the general intellectual public. While the vice of the school-bound scholar is pedantry or a monopolistic outlook, the vice of the scholar oriented to the general public may be carelessness or flippancy. The public at large may be indiscriminate, its judgments vacillating and unreliable.

There is thus no escape, if one wants to make use of the freedom which in principle teaching allows, from having to test one's work against the inevitable subjectivity of one's own judgments. Such subjectivity can, of course, be tempered by reading. It can also be tempered, and the life of the mind rendered more vivacious and stimulating, when the college teacher learns how to find his true colleagues. They will not necessarily be found in his own department or a closely related one: a man teaching sociology may find, for instance, that one of the historians in his university is closer to his way of thinking and feeling than are the other sociologists, the cultural anthropologists, or the social psychologists.

Social Scientists

In many colleges social science is today, for complex reasons, a particularly exciting undergraduate field, for students bring to it concerns about themselves, their society, and other societies that in an earlier day they might have brought—and in some places still do bring—to their courses in literature, religion, and philosophy. Sociology, anthropology, and some sorts of psychology thus have become, for the time being and at some universities, areas of intellectual openness and vitality; and those who teach in these fields, far from having to persuade students to be interested in what we the faculty have to offer, are almost in the converse position of having to fight them off! This is not an unmixed blessing, of course. Nothing in teaching is. And if some professors in other fields envy the appeal sociology has for students and indeed regard it as somehow illicit as well as unfair competition, teachers like myself may occasionally wish they taught a subject that was neatly bounded, one that had no connection either with their own emotional life or with that of their students, and one that appealed only to a small, select group that could pursue its problems in genteel quiet and amicability.

For assuredly one of the great problems for some social scientists, even more than for other academic people, is that it is almost impossible to stop working. Everything is grist for the mill, or at least is thought to be; so that if I attend a party, people think I am observing them even when I am not, and if they meet me on a plane they ask me whether or not I think they are upper-middlebrows, or things of the same sort. I know that psychiatrists have much the same problem. But it is a problem less of the expectations of others than of one's finding ways to stop work, to stop being analytical, to stop psychologizing and sociologizing all of life. Since many people are brought into social science as a way of sublimating their aggressions against themselves and others, this danger is quite general. And yet at the same time it means that the social scientist does not lead the compartmentalized life of most Americans for whom things do not connect with one another.

I should say that, for me and for many others in my field, the task and opportunity of making connections—that is, of synthesis—is what we find most exciting in our work as social scientists. We try to bring together and connect things which are often kept apart, whether these be disciplines or orders of data or metaphorical ways of thinking. In our relations with students, we can try to establish connections with where they are now, to find out how they react to material and see their world. Some of us (like some novelists) find that only material that degrades man is really "real," and we seek to rub our students' noses and sometimes our own in human sadism, haplessness, and fatalism. Social science in general has revolted against the genteel tradition, against any effort to set up some orders of data as hierarchically superior to others, so that we insist that popular songs may be as worthy of study, if not as pleasant, as Mozart arias. But such reversals of traditional hierarchies and snobberies seem to me no longer such a fighting issue; the battle has largely been won. Social science at its best seems to me one of the humanities, one of the most humane and liberating ways of approaching the human condition. Thus sociology today is a kind of intellectual switchboard for some of its practitioners, with ties to the humanities, to history, to contemporary life.

On inspection, each of these ties will turn out to be a tie not only to data but also to individuals who deal with the data in one way or another. The generalizing sociologist therefore is apt to have contact with many other academic disciplines. But this is only a special case of the general situation of the professor who, if he is at all thoughtful and sensitive and wide-ranging, will find himself dealing with many complex and conflicting constituencies. He will have, first of all, the constituency of his own discipline or subdiscipline, and this is the easiest and safest to which to listen. It gives the most security and in general the most unequivocal rewards. But the "life" of a discipline is always at some remove from the life around it. This is a creative tension at best, for as I indicated earlier, it is not a good thing for the academy to become too much like the rest of the country, even if the rest of the country prospers by becoming more like the academy. The discipline must have its own standards, its own abstractions; and yet, at the same time, the resort to life itself. The dialectic here is not so very different from the dialectic between art and life, and it is subject to many of the same plagiarisms and pitfalls.

Then there is the constituency of the student body. There are some teachers who devote themselves entirely to this constituency. One can imagine a society in which such devotion would not be problematic; it would be a society more cooperative than ours, more humane, and one less adulatory of youth. But, as things are, the college professor who devotes himself uninterruptedly to his students, unless he is an extraordinary person, is likely to end up either very dry or very damp. As he ages, while of course his students do not, he may have less and less to give them in terms of intellectual leadership and grasp of the world we live in, while the warmth and friendship he continues to offer his students, though always desirable, provide benefits which many parents and other adults in our society, eager to be pals with the young, also furnish. In my judgment, based upon and perhaps biased by my own preference, close contact with students requires intermissions of colleagueship with one's own peers.

However, in the major universities that are also the major centers of graduate training, there is little danger that students will overestimate the virtues of teaching and underestimate those of research. The fashionable tendencies all run the other way. In the social sciences as in the natural sciences, there has been since World War II an immense development of research institutes where no or very little teaching is done. Much is lost here in variety and change of pace, and in the pleasure of teaching speculative, sensitive young people. Moreover, there is the risk that the research in the institutes, as in many graduate schools, will not be stimulated and reshaped by contact with young minds not yet in the groove of the discipline.

Still another constituency is the outside world, which, for the reasons I also indicated when I spoke of the infiltration of academic values into business, is eager for guidance from the academy (as increasingly are governments and voluntary organizations).

Despite all I have said about infiltration, there are of course still considerable differences between life inside and life outside a university. One of the differences that I have found in my own limited experience concerns the quality of personal relations. Competition in academic life has an especially biting quality, in some measure because we try to set our own goals and never have a balance sheet to back us up—a situation which, as in other creative pursuits, leads often to malice, envy, neurosis, and self-doubt. I would certainly warn anyone not to enter teaching if he plans to do so because he thinks the people in it are so nice! And conversely, certainly no one should avoid business because he has been persuaded by recent novels that business is a jungle.

The last dialectic I want to mention—the constituency that is perhaps most difficult to handle—is that between one's own solitary mind and the minds of one's closest intellectual allies. Indeed, in a large university there are apt to be so many colleagues in one's field, however defined, as to constitute a burden as well as a benefit; and many professors spend all their time on a plane of flat contemporaneity, keeping up with the news and gossip of the profession, with what is euphemistically termed "the literature," and with the other busy-work of which we pretend to complain. Some of this is probably inescapable, but one also needs "colleagues" with whom one is in touch vicariously, through books and other monuments of their work. At best, one may also hope for vicarious colleagues who are not yet in existence, who will be called into being by one's work as this is transmitted through readers and students. In my own case, at least, I hope such colleagues would not be disciples, bound by dogma, but independent minds, in and out of the universities, who enjoy the carrying on of scholarly traditions, the play of ideas, and the uncovering of an ever less distorted picture of man's nature.

I have tried to draw attention to some of the pitfalls and dilemmas of teaching, as I might point them out to a prospective college teacher. I have wanted to make sure that in choosing this calling, he should not allow himself to be trapped by his own teachers, bribed by fashion, or put off by fear. I have wanted to emphasize that the intellectual life is not necessarily pursued either in a more serious or more carefree way in the academic world; "academic" enclaves can be found in business and in the professions, and it goes without saying that the number of hours one puts in at a particular activity matters much less than the intensity or quality of that activity. Teaching is merely one possible way of achieving what Erich Fromm terms "relatedness"; and obviously the problems of being a man transcend those of being any particular sort of professional man.

❊ *Alfred North Whitehead*

9

The Aims of Education

In his famous essay called "The Aims of Education," delivered as his presidential speech to the Mathematical Association of England in 1916, Alfred North Whitehead addressed himself ostensibly to the teaching of mathematics in the British schools. But, as he explained in the introduction to a book which includes this essay among others, his remarks referred to education in general, not only in England but also in the United States—"the general principles apply equally to both countries." The essay, repub-*

* *The Aims of Education and Other Essays,* copyrighted 1929 by The Macmillan Company. This essay is reprinted with the permission of the publisher.

lished here in part (omitting some of the specific discussion of mathematics), still speaks so clearly and wisely on the educational problems of our day that it makes a fitting conclusion to this book.

Culture is activity of thought, and receptiveness to beauty and humane feeling. Scraps of information have nothing to do with it. A merely well-informed man is the most useless bore on God's earth. What we should aim at producing is men who possess both culture and expert knowledge in some special direction. Their expert knowledge will give them the ground to start from, and their culture will lead them as deep as philosophy and as high as art. We have to remember that the valuable intellectual development is self-development, and that it mostly takes place between the ages of sixteen and thirty. As to training, the most important part is given by mothers before the age of twelve. A saying due to Archbishop Temple illustrates my meaning. Surprise was expressed at the success in after-life of a man, who as a boy at Rugby had been somewhat undistinguished. He answered, "It is not what they are at eighteen, it is what they become afterwards that matters."

In training a child to activity of thought, above all things we must beware of what I will call "inert ideas"—that is to say, ideas that are merely received into the mind without being utilised, or tested, or thrown into fresh combinations.

In the history of education, the most striking phenomenon is that schools of learning, which at one epoch are alive with a ferment of genius, in a succeeding generation exhibit merely pedantry and routine. The reason is, that they are overladen with inert ideas. Education with inert ideas is not only useless: it is, above all things, harmful—*Corruptio optimi, pessima.* Except at rare intervals of intellectual ferment, education in the past has been radically infected with inert ideas. That is the reason why uneducated clever women, who have seen much of the world, are in middle life so much the most cultured part of the community. They have been saved from this horrible burden of inert ideas. Every intellectual revolution which has ever stirred humanity into greatness has been a passionate protest against inert ideas. Then, alas, with pathetic ignorance of human psychology, it has proceeded by some educational scheme to bind humanity afresh with inert ideas of its own fashioning.

Let us now ask how in our system of education we are to guard against this mental dryrot. We enunciate two educational commandments, "Do not teach too many subjects," and again, "What you teach, teach thoroughly."

The result of teaching small parts of a large number of subjects is the passive reception of disconnected ideas, not illumined with any spark of vitality. Let the main ideas which are introduced into a child's education be few and important, and let them be thrown into every combination possible. The child should make them his own, and should understand their application here and now in the circumstances of his actual life. From the very beginning of his education, the child should experience the joy of discovery. The discovery which he has to make is that general ideas give an understanding of that stream of events which pours through his life, which is his life. By understanding I mean more than a mere logical analysis, though that is included. I mean "understanding" in the sense in which it is used in the French proverb, "To understand all, is to forgive all." Pedants sneer at an education which is useful. But if education is not useful, what is it? Is it a talent, to be hidden away in a napkin? Of course, education should be useful, whatever your aim in life. It was useful to Saint Augustine and it was useful to Napoleon. It is useful, because understanding is useful.

I pass lightly over that understanding which should be given by the literary side of education. Nor do I wish to be supposed to pronounce on the relative merits of a classical or a modern curriculum. I would only remark that the understanding which we want is an understanding of an insistent present. The only use of a knowledge of the past is to equip us for the present. No more deadly harm can be done to young minds than by depreciation of the present. The present contains all that there is. It is holy ground; for it is the past, and it is the future. At the same time it must be observed that an age is no less past if it existed two hundred years ago than if it existed two thousand years ago. Do not be deceived by the pedantry of dates. The ages of Shakespeare and of Molière are no less past than are the ages of Sophocles and of Virgil. The communion of saints is a great and inspiring assemblage, but it has only one possible hall of meeting, and that is, the present; and the mere lapse of time through which any particular group of saints must travel to reach that meeting-place makes very little difference.

Passing now to the scientific and logical side of education, we remember that here also ideas which are not utilised are positively harmful. By utilising an idea, I mean relating it to that stream, compounded of sense perceptions, feelings, hopes, desires, and of mental activities adjusting thought to thought, which forms our life. I can imagine a set of beings which might fortify their souls by passively reviewing disconnected ideas. Humanity is not built that way—except perhaps some editors of newspapers.

In scientific training, the first thing to do with an idea is to prove it. But allow me for one moment to extend the meaning of "prove": I mean—to prove its worth. Now an idea is not worth much unless the propositions in which it is embodied are true. Accordingly an essential part of the proof of an idea is the proof, either by experiment or by logic, of the truth of the propositions. But it is not essential that this proof of the truth should constitute the first introduction to the idea. After all, its assertion by the authority of respectable teachers is sufficient evidence to begin with. In our first contact with a set of propositions, we commence by appreciating their importance. That is what we all do in after-life. We do not attempt, in the strict sense, to prove or to disprove anything, unless its importance makes it worthy of that honour. These two processes of proof, in the narrow sense, and of appreciation, do not require a rigid separation in time. Both can be proceeded with nearly concurrently. But in so far as either process must have the priority, it should be that of appreciation by use. . . .

We are only just realising that the art and science of education require a genius and a study of their own; and that this genius and this science are more than a bare knowledge of some branch of science or of literature. This truth was partially perceived in the past generation; and headmasters, somewhat crudely, were apt to supersede learning in their colleagues by requiring left-hand bowling and a taste for football. But culture is more than cricket, and more than football, and more than extent of knowledge.

Education is the acquisition of the art of utilisation of knowledge. This is an art very difficult to impart. Whenever a textbook is written of real educational worth, you may be quite certain that some reviewer will say that it will be difficult to teach from it. Of course it will be difficult to teach from it. If it were easy, the book ought to be burned; for it cannot be educational. In education, as elsewhere, the broad primrose path leads to a nasty place. This evil path is represented by a book or a set of lectures which will practically enable the student to learn by heart all the questions likely to be asked at the next external examination. And I may say in passing that no educational system is possible unless every question directly asked of a pupil at any examination is either framed or modified by the actual teacher of that pupil in that subject. The external assessor may report on the curriculum or on the performance of the pupils, but never should be allowed to ask the pupil a question which has not been strictly supervised by the actual teacher, or at least inspired by a long conference with him. There are a few exceptions to this rule, but they are exceptions, and could easily be allowed for under the general rule.

We now return to my previous point, that theoretical ideas should always find important applications within the pupil's curriculum. This is not an easy doctrine to apply, but a very hard one. It contains within itself the problem of keeping knowledge alive, of preventing it from becoming inert, which is the central problem of all education.

The best procedure will depend on several factors, none of which can be neglected, namely, the genius of the teacher, the intellectual type of the pupils, their prospects in life, the opportunities offered by the immediate surroundings of the school, and allied factors of this sort. It is for this reason that the uniform external examination is so deadly. We do not denounce it because we are cranks, and like denouncing established things. We are not so childish. Also, of course, such examinations have their use in testing slackness. Our reason of dislike is very definite and very practical. It kills the best part of culture. When you analyse in the light of experience the central task of education, you find that its successful accomplishment depends on a delicate adjustment of many variable factors. The reason is that we are dealing with human minds, and not with dead matter. The evocation of curiosity, of judgment, of the power of mastering a complicated tangle of circumstances, the use of theory in giving foresight in special cases—all these powers are not to be imparted by a set rule embodied in one schedule of examination subjects.

I appeal to you, as practical teachers. With good discipline, it is always possible to pump into the minds of a class a certain quantity of inert knowledge. You take a textbook and make them learn it. So far, so good. The child then knows how to solve a quadratic equation. But what is the point of teaching a child to solve a quadratic equation? There is a traditional answer to this question. It runs thus: The mind is an instrument, you first sharpen it, and then use it; the acquisition of the power of solving a quadratic equation is part of the process of sharpening the mind. Now there is just enough truth in this answer to have made it live through the ages. But for all its half-truth, it embodies a radical error which bids fair to stifle the genius of the modern world. I do not know who was first responsible for this analogy of the mind to a dead instrument. For aught I know, it may have been one of the seven wise men of Greece, or a committee of the whole lot of them. Whoever was the originator, there can be no doubt of the authority which it has acquired by the continuous approval bestowed upon it by eminent persons. But whatever its weight of authority, whatever the high approval which it can quote, I have no hesitation in denouncing it as one of the most fatal, erroneous, and dangerous conceptions ever introduced into the theory of education. The mind is never passive; it is a perpetual activity, delicate, receptive, responsive to stimulus. You cannot postpone its life until you have sharpened it. Whatever interest attaches to your subject-matter must be evoked here and now; whatever powers you are strengthening in the pupil, must be exercised here and now; whatever possibilities of mental life your teaching should impart, must be exhibited here and now. That is the golden rule of education, and a very difficult rule to follow.

The difficulty is just this: the apprehension of general ideas, intellectual habits of mind, and pleasurable interest in mental achievement can be evoked by no form of words, however accurately adjusted. All practical teachers know that education is a patient process of the mastery of details, minute by minute, hour by hour, day by day. There is no royal road to learning through an airy path of brilliant generalisations. There is a proverb about the difficulty of seeing the wood because of the trees. That difficulty is exactly the point which I am enforcing. The problem of education is to make the pupil see the wood by means of the trees.

The solution which I am urging is to eradicate the fatal disconnection of subjects which kills the vitality of our modern curriculum. There is only one subject-matter for education, and that is Life in all its manifestations. Instead of this single unity, we offer children—Algebra, from which nothing follows; Geometry, from which nothing follows; Science, from which nothing follows; History, from which nothing follows; a Couple of Languages, never mastered; and lastly, most dreary of all, Literature, represented by plays of Shakespeare, with philological notes and short analyses of plot and character to be in substance committed to memory. Can such a list be said to represent Life, as it is known in the midst of the living of it? The best that can be said of it is that it is a rapid table of contents which a deity might run over in his mind while he was thinking of creating a world, and had not yet determined how to put it together.

Let us now return to quadratic equations. We still have on hand the unanswered question. Why should children be taught their solution? Unless quadratic equations fit into a connected curriculum, of course there is no reason to teach anything about them. . . .

Quadratic equations are part of algebra, and algebra is the intellectual instrument which has been created for rendering clear the quantitative aspects of the world. There is no getting out of it. Through and through the world is infected with quantity. To talk sense is to talk in quantities. It is no use saying that the nation is large—How large? It is no use saying that radium is scarce—How scarce? You cannot evade quantity. You may fly to poetry and to music, and quantity and number will face you in your rhythms and octaves. Elegant intellects which despise the theory of quantity are but half developed. They are more to be pitied than blamed. The scraps of gibberish, which in their schooldays were taught to them in the name of algebra, deserve some contempt.

This question of the degeneration of algebra into gibberish, both in word and in fact, affords a pathetic instance of the uselessness of reforming educational schedules without a clear conception of the attributes which you wish to evoke in the living minds of the children. A few years ago there was an outcry that school algebra was in need of reform, but there was a general agreement that graphs would put everything right. So all sorts of things were extruded, and graphs were introduced. So far as I can see, with no sort of idea behind them, but just graphs. Now every examination paper has one or two questions on graphs. Personally, I am an enthusiastic adherent of graphs. But I wonder whether as yet we have gained very much.

You cannot put life into any schedule of general education unless you succeed in exhibiting its relation to some essential characteristic of all intelligent or emotional perception. It is a hard saying, but it is true; and I do not see how to make it any easier. In making these little formal alterations you are beaten by the very nature of things. You are pitted against too skilful an adversary, who will see to it that the pea is always under the other thimble.

Reformation must begin at the other end. First, you must make up your mind as to those quantitative aspects of the world which are simple enough to be introduced into general education; then a schedule of algebra should be framed which will about find its exemplification in these applications. We need not fear for our pet graphs, they will be there in plenty when we once begin to treat algebra as a serious means of studying the world. Some of the simplest applications will be found in the quantities which occur in the simplest study of society. The curves of history are more vivid and more informing than the dry catalogues of names and dates which comprise the greater part of that arid school study. What purpose is effected by a catalogue of undistinguished kings and queens? Tom, Dick or Harry, they are all dead. General resurrections are failures, and are better postponed. The quantitative flux of the forces of modern society is capable of very simple exhibition. Meanwhile, the ideas of the variable, of the function, of rate of change, of equations and their solution, of elimination, are being studied as an abstract science for their own sake. Not, of course, in the pompous phrases with which I am alluding to them here, but with that iteration of simple special cases proper to teaching.

If this course be followed, the route from Chaucer to the Black Death, from the Black Death to modern Labour troubles, will connect the tales of the mediaeval pilgrims with the abstract science of algebra, both yielding diverse aspects of that single theme, Life. I know what most of you are thinking at this point. It is that the exact course which I have sketched out is not the particular one which you would have chosen, or even see how to work. I quite agree. I am not claiming that I could do it myself. But your objection is the precise reason why a common external examination system is fatal to education. The process of exhibiting the applications of knowledge must, for its success, essentially depend on the character of the pupils and the genius of the teacher. Of course I have left out the easiest applications with which most of us are more at home. I mean the quantitative sides of sciences, such as mechanics and physics. . . .

Finally, if you are teaching pupils for some general examination, the problem of sound teaching is greatly complicated. Have you ever noticed the zig-zag moulding round a Norman arch? The ancient work is beautiful, the modern work is hideous. The reason is, that the modern work is done to exact measure, the ancient work is varied according to the idiosyncrasy of the workman. Here it is crowded, and there it is expanded. Now the essence of getting pupils through examinations is to give equal weight to all parts of the schedule. But mankind is naturally specialist. One man sees a whole subject, where another can find only a few detached examples. I know that it seems contradictory to allow for specialism in a curriculum especially designed for a broad culture. Without contradictions the world would be simpler and perhaps duller. But I am certain that in education wherever you exclude specialism you destroy life. . . .

Fortunately, the specialist side of education presents an easier problem than does the provision of a general culture. For this there are many reasons. One is that many of the principles of procedure to be observed are the same in both cases, and it is unnecessary to recapitulate. Another reason is that specialist training takes place—or should take place—at a more advanced stage of the pupil's course, and thus there is easier material to work on. But undoubtedly the chief reason is that the specialist study is normally a study of peculiar interest to the student. He is studying it because, for some reason, he wants to know it. This makes all the difference. The general culture is designed to foster an activity of mind; the specialist course utilises this activity. But it does not do to lay too much stress on these neat antitheses. As we have already seen, in the general course foci of special interest will arise; and similarly in the special study, the external connections of the subject drag thought outwards.

Again, there is not one course of study which merely gives general culture, and another which gives special knowledge. The subjects pursued for the sake of a general education are special subjects specially studied; and, on the other hand, one of the ways of encouraging general mental activity is to foster a special devotion. You may not divide the seamless coat of learning. What education has to impart is an intimate sense for the power of ideas, for the beauty of ideas, and for the structure of ideas, together with a particular body of knowledge which has peculiar reference to the life of the being possessing it.

The appreciation of the structure of ideas is that side of a cultured mind which can only grow under the influence of a special study. I mean that eye for the whole chessboard, for the bearing of one set of ideas on another. Nothing but a special study can give any appreciation for the exact formulation of general ideas, for their relations when formulated, for their service in the comprehension of life. A mind so disciplined should be both more abstract and more concrete. It has been trained in the comprehension of abstract thought and in the analysis of facts.

Finally, there should grow the most austere of all mental qualities: I mean the sense for style. It is an aesthetic sense, based on admiration for the direct attainment of a foreseen end, simply and without waste. Style in art, style in literature, style in science, style in logic, style in practical execution have fundamentally the same aesthetic qualities, namely, attainment and restraint. The love of a subject in itself and for itself, where it is not the sleepy pleasure of pacing a mental quarter-deck, is the love of style as manifested in that study.

Here we are brought back to that position from which we started, the utility of education. Style, in its finest sense, is the last acquirement of the educated mind; it is also the most useful. It pervades the whole being. The administrator with a sense for style hates waste; the engineer with a sense for style economises his material; the artisan with a sense for style prefers good work. Style is the ultimate morality of mind.

But above style and above knowledge, there is something, a vague shape like fate above the Greek gods. That something is power. Style is the fashioning of power, the restraining of power. But, after all, the power of attainment of the desired end is fundamental. The first thing is to get there. Do not bother about your style, but solve your problem, justify the ways of God to man, administer your province, or do whatever else is set before you.

Where, then, does style help? In this: with style the end is attained without side issues, without raising undesirable inflammations. With style you attain your end and nothing but your end. With style the effect of your activity is calculable, and foresight is the last gift of gods to men. With style your power is increased, for your mind is not distracted with irrelevancies, and you are more likely to attain your object. Now style is the exclusive privilege of the expert. Whoever heard of the style of an amateur painter, of the style of an amateur poet? Style is always the product of specialist study, the peculiar contribution of specialism to culture.

English education in its present phase suffers from a lack of a definite aim, and from an external machinery which kills its vitality. Hitherto in this address I have been considering the aims which should govern education. In this respect England halts between two opinions. It has not decided whether to produce amateurs or experts. The profound change in the world which the nineteenth century has produced is that the growth of knowledge has given foresight. The amateur is essentially a man with appreciation and with immense versatility in mastering a given routine. But he lacks the foresight which comes from special knowledge. The object of this address is to suggest how to produce the expert without loss of the essential virtues of the amateur. The machinery of our secondary education is rigid where it should be yielding, and lax where it should be rigid. Every school is bound on pain of extinction to train its boys for a small set of definite examinations. No headmaster has a free hand to develop his general education or his specialist studies in accordance with the opportunities of his school, which are created by its staff, its environment, its class of boys, and its endowments. I suggest that no system of external tests which aims primarily at examining individual scholars can result in anything but educational waste.

Primarily it is the schools and not the scholars which should be inspected. Each school should grant its own leaving certificates, based on its own curriculum. The standards of these schools should be sampled and corrected.

But the first requisite for educational reform is the school as a unit, with its approved curriculum based on its own needs, and evolved by its own staff. If we fail to secure that, we simply fall from one formalism into another, from one dunghill of inert ideas into another.

In stating that the school is the true educational unit in any national system for the safeguarding of efficiency, I have conceived the alternative system as being the external examination of the individual scholar. But every Scylla is faced by its Charybdis—or, in more homely language, there is a ditch on both sides of the road. It will be equally fatal to education if we fall into the hands of a supervising department which is under the impression that it can divide all schools into two or three rigid categories, each type being forced to adopt a rigid curriculum. When I say that the school is the educational unit, I mean exactly what I say, no larger unit, no smaller unit. Each school must have the claim to be considered in relation to its special circumstances. The classifying of schools for some purposes is necessary. But no absolutely rigid curriculum, not modified by its own staff, should be permissible. Exactly the same principles apply, with the proper modifications, to universities and to technical colleges.

When one considers in its length and in its breadth the importance of this question of the education of a nation's young, the broken lives, the defeated hopes, the national failures, which result from the frivolous inertia with which it is treated, it is difficult to restrain within oneself a savage rage. In the conditions of modern life the rule is absolute: the race which does not value trained intelligence is doomed. Not all your heroism, not all your social charm, not all your wit, not all your victories on land or at sea, can move back the finger of fate. Today we maintain ourselves. Tomorrow science will have moved forward yet one more step, and there will be no appeal from the judgment which will then be pronounced on the uneducated.

We can be content with no less than the old summary of educational ideal which has been current at any time from the dawn of our civilisation. The essence of education is that it be religious.

Pray, what is religious education?

A religious education is an education which inculcates duty and reverence. Duty arises from our potential control over the course of events. Where attainable knowledge could have changed the issue, ignorance has the guilt of vice. And the foundation of reverence is this perception, that the present holds within itself the complete sum of existence, backwards and forwards, that whole amplitude of time which is eternity.

DATE D
2 4 NOV. 1980